Oxford International Primary

W0036498

6

Science

Student Book

Deborah Roberts
Terry Hudson

Alan Haigh
Geraldine Shaw

Language consultants:
John McMahon
Liz McMahon

OXFORD

OXFORD
UNIVERSITY PRESS

Great Clarendon Street, Oxford, OX2 6DP, United Kingdom

Oxford University Press is a department of the University of Oxford. It furthers the University's objective of excellence in research, scholarship, and education by publishing worldwide. Oxford is a registered trade mark of Oxford University Press in the UK and in certain other countries.

© Deborah Roberts, Terry Hudson, Alan Haigh and Geraldine Shaw 2021

The moral rights of the authors have been asserted.

First published in 2014

All rights reserved. No part of this publication may be reproduced, stored in a retrieval system, or transmitted, in any form or by any means, without the prior permission in writing of Oxford University Press, or as expressly permitted by law, by licence or under terms agreed with the appropriate reprographics rights organization. Enquiries concerning reproduction outside the scope of the above should be sent to the Rights Department, Oxford University Press, at the address above.

You must not circulate this work in any other form and you must impose this same condition on any acquirer.

British Library Cataloguing in Publication Data

Data available

ISBN 978-1-382006590

7 9 10 8 6

Paper used in the production of this book is a natural, recyclable product made from wood grown in sustainable forests. The manufacturing process conforms to the environmental regulations of the country of origin.

Printed in China by Golden Cup

Acknowledgements

The publisher and authors would like to thank the following for permission to use photographs and other copyright material:

Cover:. Artwork by Blindsalida. **Photos: p10(l):** PhiveT/Alamy Stock Photo; **p10(r):** Gtranquillity/Shutterstock; **p12:** hoangminh1904/Shutterstock; **p13(l):** Cultura Creative RF/Alamy Stock Photo; **p13(r):** Martin Shields/Alamy Stock Photo; **p14–15:** Vibrant Image Studio/Shutterstock; **p15(t):** Sabena Jane Blackbird/Alamy Stock Photo; **p15(m):** Rich Carey/Shutterstock; **p16:** Jochen Tack/Alamy Stock Photo; **p16(bl):** VectorMine/Shutterstock; **p17:** Harry Collins Photography/Shutterstock; **p18(tl):** critterbiz/Shutterstock; **p18(a):** clayton harrison/Shutterstock; **p18(b):** Jim Cumming/Shutterstock; **p18(c):** katatonia82/Shutterstock; **p18(d):** Ondrej Prosicky/Shutterstock; **p18(e):**matushaban/Shutterstock; **p18(f):** Robert Chao/Shutterstock; **p19:** Pictorial Press Ltd/Alamy Stock Photo; **p20(tl):** Fish Without Panties/Shutterstock; **p20(tr):** marilyn barbone/Shutterstock; **p20(ml):** Modella/Shutterstock; **p20(m):**5r82/Shutterstock; **p20(mr):** Viewnature/Shutterstock; **p20(br):** rustamank/Shutterstock; **p21:** Andrii Vodolazhskyi/Shutterstock; **p22(tl):** Images & Stories/Alamy Stock Photo; **p22(bl):** nrey/Shutterstock; **p22(br):** Peter Gudella/Shutterstock; **p23:** Rashid Valitov/Shutterstock; **p24(t), p144:** leungchopan/Shutterstock; **p24(b):** Teo Tarras/Shutterstock; **p25:** Belova Daria/Shutterstock; **p26(t):** Toa55/Shutterstock; **p26(br):** wcjohnston/iStock/Getty Images; **p26(bl):** Science History Images/Alamy Stock Photo; **p28(t):** zstock/Shutterstock; **p28(b):** BMJ/Shutterstock; **p30(t):** Lou Linwei/Alamy Stock Photo; **p30(b):** Art Directors & TRIP/Alamy Stock Photo; **p31(t):** Huguette Roe/Shutterstock; **p31(b):** Joel W. Rogers/Getty Images; **p32:** leungchopan/Shutterstock; **p33(t):** manfredxy/Shutterstock; **p33(b):** Shipov Oleg/Shutterstock;

p37(tr): Rich Carey/Shutterstock; **p37(br):** photopixel/Shutterstock; **p38:** Wessel du Plooy/Shutterstock; **p39(tr):** RG Images/STOCK4B-RF/OUP; **p39(br):** Richard Coombs/Alamy Stock Photo; **p42–43:** James Porcini/Cultura/Getty Images; **p49(l):** Lou Linwei/Alamy Stock Photo; **p49(r):** Nerthuz/Shutterstock; **p59:** Avpics/Alamy Stock Photo; **p60(l):**patanasak/iStockphoto; **p60(r):** alexionas/iStockphoto; **p62:** aapsky/Shutterstock; **p64:** Jim Mone/Associated Press; **p65:**ashok india/Shutterstock; **p68–69:** EyeEm/Alamy Stock Photo; **p69(tl):** Vadim Petrakov/Shutterstock; **p70(m):** nigel baker photography/Shutterstock; **p70(l):** WAYHOME studio/Shutterstock; **p70(r):** Acon Cheng/Shutterstock; **p72:** rck_953/Shutterstock; **p73:** chomplearn/Shutterstock; **p75(tr):** godrick/Shutterstock; **p75(bl):** Castleski/Shutterstock; **p76(a):**Tom Wang/Shutterstock; **p76(b):** ldambies/Shutterstock; **p76(d):** Fabio Alcini/Shutterstock; **p76(f):** Artazum/Shutterstock; **p76(g):** Voyagerix/Shutterstock; **p76(c):** Westend61 GmbH/Alamy Stock Photo; **p76(e):** Gaf_Lila/Shutterstock; **p78:** Image Source/OUP; **p79:** Photodisc/OUP; p80: Design Pics Inc/Alamy Stock Photo; **p82(t):** MJTH/Shutterstock; **p82(b):** James Woodson/Getty Images; **p84:**MilanB/Shutterstock; **p86(a):** Vladitto/Shutterstock; **p86(b):** Franck Boston/Shutterstock; **p86(c):** Chones/Shutterstock; **p86(d):** David Butow/Corbis Historical/Getty Images; **p86(e):** Olga Miltsova/Shutterstock; **p86(f):** Maridav/Shutterstock; **p86(g):** zentilia/Shutterstock; **p88:** FreshPaint/Shutterstock; **p89:** Photodisc/OUP; **p90:** Photobank/Shutterstock; **p94:** Pekka Parviainen/Science Photo Library; **p95:** Maskot/Getty Images; **p96:** Ingram/OUP; **p98:** Heintje Joseph T. Lee/Shutterstock; **p99(tr):** Ocean/OUP; **p99(bl):** Viktor1/Shutterstock; **p99(br):** Alexander Kalina/Shutterstock; **p100(t):** NASA, NOAA NGDC, Suomi-NPP, Earth Observatory; **p100(b):** Popperfoto/Getty Images; **p104–105:** arhip4/Shutterstock; **p105(b):**skaljac/Shutterstock; **p105(t):** fStop/OUP; **p106:** Flegere/Shutterstock; **p107:** negaprion/iStockphoto; **p108:**yevgeniy11/Shutterstock; **p109:** Trevor Clifford Photography/Science Photo Library; **p110(tr):** Martyn F. Chillmaid/Science Photo Library; **p111:** Sunshine boy/Shutterstock; **p110(tl):** asadykov/Shutterstock; **p113:** Rawpixel.com/Shutterstock; **p114(t):**Monster Ztudio/Shutterstock; **p114(bl):** Natee Photo/Shutterstock; **p114(br):** Shine lal/Shutterstock; **p116:** Tatiana Popova/Shutterstock; **p120(t):** The Photo Works/Alamy Stock Photo; **p120(b):** sciencephotos/Alamy Stock Photo; **p121:** tristan tan/Shutterstock; **p126–127:** Stu Porter/Shutterstock; **p127(t), p146(t):** Natural History Museum, London/Science Photo Library; **p127(m):** IanRedding/Shutterstock; **p128, p145:** Snowshill/Shutterstock; **p129(t):** releon8211/Shutterstock; **p129(ml):** Photoongraphy/Shutterstock; **p129(m):** UMB-O/Shutterstock; **p129(mr):** Galyna Andrushko/Shutterstock; **p130:**Dinoton/Shutterstock; **p132(a):** muroPhotographer/Shutterstock; **p132(b):** Javi Roces/Shutterstock; **p132(c):** Eric Isselee/Shutterstock; **p132(d):** SeraphP/Shutterstock; **p132(e):** Abramova Kseniya/Shutterstock; **p132(f):** Johann Hinrichs/Alamy Stock Photo; **p133(t):** Zuzha/Shutterstock; **p133(b):** Henrik Larsson/Shutterstock; **p134(a):**moosehenderson/Shutterstock; **p134(b):** Vudhikrai/Shutterstock; **p134(c):** John Arnold/Shutterstock; **p134(d):** Pichugin Dmitry/Shutterstock; **p134(e):** Matt Jeppson/Shutterstock; **p135:** Nigel Cattlin/Alamy Stock Photo; **p136(tl):** Alen thien/Shutterstock; **p136(tr):** Steve Byland/Shutterstock; **p136(mr), p140(b):** outdoorsman/Shutterstock; **p137:** Anastasiia Malinich/Shutterstock; **p138(t):** IanRedding/Shutterstock; **p138(b):** Sanatana/Shutterstock; **p146(b):** Somogyi Laszlo/Shutterstock; **p147:** FatCamera/E+/Getty Images; **p148:** Elnur/Shutterstock.

Artwork by Q2A Media Services Pvt. Ltd.

Every effort has been made to contact copyright holders of material reproduced in this book. Any omissions will be rectified in subsequent printings if notice is given to the publisher.

Contents

Contents

How to Use this Book

This Student Book for *Oxford International Primary Science* forms part of your science lessons for this year. Your teacher will introduce the ideas through whole-class activities, then you will explore them in more detail using this book, before all coming back together to discuss what you have learned. Find out more at: www.oxfordprimary.com/international-science

Structure of the book

This book is divided into five units plus an introduction called *Being a Good Scientist* and a picture Glossary:

Being a Good Scientist
Unit 1 Classification and Habitats
Unit 2 Organs and Systems
Unit 3 The Way We See Things
Unit 4 Building Electrical Circuits
Unit 5 Adaptation and Inherited Characteristics
Glossary

Each unit covers a different strand of science. You will need a science notebook to write in and to record your investigation results and conclusions.

Being a good scientist

To be a good scientist you need to be curious and ask questions. This section will help you think about how to develop your scientific skills to work like a scientist.

What you will find in each unit

There are three types of lessons:
Wow introduces each unit's scientific ideas and key words. It tells you what you will learn in the unit and lets you discuss what you already know.
Focused lessons cover the scientific knowledge and skills you need to learn this year.
In **What have I learned** you review your understanding and show your teacher what you have learned about the unit.

What you will find in the lessons

Although each lesson is unique, they have common features:

The words on the Wow pages are included in the picture glossary at the back of the book. You can add your own notes for each word.

Key words
alkali
pH scale
pollution
waste

Gives you the key words for the lesson.

In this lesson you will learn how water is transported and excreted. Tells you what you will learn in the lesson.

Questions to help you talk to each other and share ideas about the science you are learning and the investigations you do.

Practical and research activities to investigate and report on science topics. Sometimes your teacher will ask you to use different equipment, which is available in school. They may also ask you to carry out a test in a different way, to make sure you are safe.

Stretch zone ➔ Challenges you to take your learning further.

Key idea Summarises what you have learned.

Additional features

Think back Reminds you what has been covered before.

Science fact Interesting and amazing science facts.

Highlights the skills needed to be a good scientist.

Important notes about how to stay safe.

Teacher's Guide

There is a Teacher's Guide to help your teacher to work out the resources needed and to offer alternative activities and approaches.

Workbook

At the bottom of each page in this book is a link to a Workbook, where you can record your work and get extra practice to do in your lesson or at home.

Being a Good Scientist

As you know, science is the study of the world around us. You will have found out that to be a good scientist you need to be curious and ask questions. This section will help you think about how to build on your scientific skills to plan and carry out more complicated investigations.

Your work as a scientist this year will allow you to develop further your scientific skills. You will make more detailed predictions and observe patterns in your results. Having detected patterns in data, you will need to decide if these are the result of your investigation or simply happened by chance. You will also need to decide if your results were accurate and valid. You will have to think more deeply about how living and non-living things are classified. You will also be expected to test your own ideas and use scientific evidence.

This diagram shows the steps you can take to plan and carry out investigations like a scientist.

Start here: Asking questions
- Exploring ideas and thinking of questions to investigate.

Making a prediction
- What do I already know that will help me to decide what is likely to happen?

Planning
- Which is the best type of test to carry out?
- What are the variables?
- Which need to be controlled and why?
- How can I find out more before I start?
- Which secondary sources shall I use?
- What equipment do I need?

Making observations
- What observations and measurements should I make?
- How long should I make observations and measurements for?
- When do I repeat measurements?
- What equipment should I use?

Recording findings
- What is the best way to record data?
- Should I present the data in tables, scatter graphs, bar graphs or line graphs?
- Should I use scientific diagrams and labels?
- Is a classification key useful?
- Can I record into a table?
- Would photographs or film be helpful?
- Would a written description work well?

Drawing conclusions
- Does the data show that the research question has been answered?
- Are further tests needed?
- Does the evidence support my ideas?
- Do I trust my results?
- How can I improve the investigation?

Presenting ideas
- Have I used scientific language and illustrations?
- Should I present my ideas by speaking, by writing or by using displays and computer presentations? Would a model help?
- Does my work lead to other questions to study?

Asking questions

You have been encouraged to start your investigation questions with words such as 'which', 'what', 'why', 'how', 'do' and 'does'. This can help to lead you towards planning an investigation or carrying out research that will have a clear answer. The better you are at forming a question, the easier you will find it to plan and carry out investigations. Different types of questions are used in different situations.

Finding out what is happening: verification questions

These questions are designed to help you to collect data to find something out about a situation. You don't need to know anything about it before your investigation. For example:

- What happens when a ray of light hits a mirror?
- Do more batteries make the bulbs in a circuit brighter?

Answers to these questions will help you to build your knowledge, and the questions will lead you towards the type of investigation to carry out.

Select one of the questions. Talk about what type of investigation you could carry out to answer the question.

Finding out why things happen: theory questions

These questions need you to have some prior knowledge of the subject. The question also means you have to explain WHY something happened. For example:

- Why do shadows change in size when the light source is moved?
- Why do heart rates and breathing rates increase during exercise?

How will we measure if the bulbs change in brightness? Does the way the bulbs are arranged make a difference?

Experimental questions

These questions grow from your prior knowledge of the topic; they need an explanation and they are testable. In other words, other people can test your answer to see if they agree. For example:

- Does adding more volts to a circuit make the bulbs shine brighter?
- Does light travel in a straight line?

Read the three questions below. Decide whether they are verification, theory or experimental questions. Talk about how you would plan to answer the questions.

1 If an extra bulb is added to a circuit, will the other bulbs be brighter or dimmer?
2 How are animals adapted to living in dry countries?
3 Why can we not see objects when a room is dark?

The fact that answering a question can lead onto others is why the investigative process is shown as a cycle.

Making a prediction

When answering a research or investigation question you should make a prediction.

This is based on what you already know about a topic. Scientists are usually confident about what will happen in an investigation. They may have done similar investigations before or read about similar work elsewhere.

Use what you know about electricity to help you think about this question.

What would you observe if you added two cells (batteries) to a circuit that already had one cell and three bulbs? Do you think the bulbs would be brighter or dimmer? What did you think about to help you decide?

As a scientist, you draw on your previous experiences to help. You think about when you have created simple circuits. This makes your prediction much better than a guess. It is based on scientific knowledge and evidence.

Scientists may use **models** and **diagrams** to represent objects and systems. These help scientists explain and think about scientific ideas that are not visible or unknown. Scientists can then use their models and diagrams to make predictions or to explain observations.

Remember: a prediction can be shown to be incorrect. An investigation, no matter how often it is repeated, may show that your original prediction cannot be the correct answer.

What would you do if your prediction is shown by your investigation to be incorrect? Use books and the internet to find out some examples of science theories that have now been proved to be incorrect.

Planning

It is vital that scientists plan what they are going to do. They discuss their plan to check it will work. A good scientist will also research the topic to find out as much as they can. They use secondary sources.

A secondary source is any source that gives you information you have not found out for yourself. Examples are written information in books and on the internet, talks from people who did the original work, documentaries, journals, magazines or newspapers.

Use secondary sources when you can but be careful. Some can be trusted more than others. Try to use more than one source of information and check it by doing investigations yourself when you can.

Remember: scientists think carefully about the equipment they will need. They make a list and make sure everything is available before they start an investigation.

Different types of test

Descriptive investigations

This is when you observe something over time and describe what happens. Every time you survey plants or animals you have been doing a descriptive investigation. You do not need to know anything about the topic and you do not need a prediction. You are recording what you see and then making sense of it. You will end up identifying, observing, listing and describing what you see.

Discuss a descriptive test you have carried out. What were you observing?

Comparative investigations

This is when you observe or measure at different times or in different places to compare data to see if there are any changes or differences.

You will be encouraged to set up what are called comparative tests. This is when you design an investigation to compare different things. For example, you could compare the size and shape of shadows with different light sources.

Experimental investigations

This is when you will be designing a fair test. This means you will have to decide on which factor or variable you will alter, which you will measure, and which you will control. The investigation is set up to gather data that supports or does not support a causal relationship. This means we are investigating if changing X causes or makes Y change.

The types of variable are described below:

- **Independent variable** (sometimes called the manipulative variable) – this is what you change on purpose in an investigation.

- **Dependent variable** (sometimes called the response variable) – this is what changes during the investigation because you have altered the independent variable. It is what you measure.

- **Control variables** (sometimes called constants) – these are the variables you keep the same during an investigation.

Study the picture. Discuss and identify the independent, dependent and control variables for this investigation. What causal effect are the students studying? What would your prediction be?

As you have found in your earlier work surveys of habitats also need to be fair. You should survey the same amount of ground so you can do a fair comparison with other areas. If you surveyed a large area in one place and a small area in another place, then it would not be a fair test: you could not fairly compare which has the most plants or insects. When you survey people, for example about what they eat, you should also try to include the same number of people every time.

Science fact

Scientists sometimes give a suggested answer to an investigation. This is called a hypothesis. If other scientists test this and they all agree, it then becomes a theory. In time, a theory that does not change can become a law of science.

Sometimes it is not possible to plan an investigation to answer your questions. If you want to explore forces that are too strong for you to investigate, you cannot carry out a test but you can use secondary sources.

What are secondary sources? List the times you have used them to find out about a topic.

Making observations

Scientists use their observation skills during investigations.

> What observations and measurements would the students investigating exercise and pulse rate be carrying out? Write a list.

During the planning stage you will decide which observations and measurements you need to make. This will depend on the type of investigation you are carrying out.

With surveys, this may involve counting different living things and observing what they look like to help with identification.

With experimental investigations, this can involve measuring the time taken for something to happen, the height that something has grown, the temperature of a material, or the number of grams of something.

Scientists also decide on the best place to carry out observations. They think about the equipment they need, the safety measures that need to be followed, and the reason for the investigation. For example, a survey of animals in a habitat is carried out in a particular outdoor location and a chemistry experiment is usually carried out in a laboratory.

Scientists use devices such as computers, data loggers and other devices, such as smartphones and electronic scales, to help them to take accurate measurements.

Science fact

Scientists use standard units to record their results. These units have been agreed throughout the world so all scientists can compare their work. The standard unit for length is the metre or kilometre.

> Which standard units would you use to measure: a) temperature, b) distance between villages, c) the amount of flour needed in a recipe?

Good scientists repeat measurements. This is to make sure they have not made any mistakes. They can then calculate a mean average for their readings. The example below shows the results of a light reflection investigation.

Angle of incidence (light reaching the mirror) (degrees)	Angle of reflection (light leaving the mirror) (degrees)			
	Measurement 1	Measurement 2	Measurement 3	Mean (average)
20	19	21	20	
30	32	29	29	
40	40	40	40	

> What is the average reading for each angle of reflection? Why is it useful to not just take the first reading? List two reasons why the result may vary every time you do this investigation.

Remember: in some investigations you may use a key to help you to identify living things and objects.

Recording findings

As part of the planning process, scientists think about the best way to record their results. They might decide to use a table or labelled diagrams. They could take photographs or film what is happening. The main thing to think about is:

> How can I record results so they help me to see patterns or to sort things into groups?

You will need to use your results to draw conclusions. This is the next part of the investigation process. If you do not record your results carefully, you may not be able to make the most sensible conclusions.

Tables

You will often record your results in a table.

> Design a table to record the shoe sizes of six people in your class.

Charts/graphs

You have presented your results as bar and line charts or graphs, like the ones below.

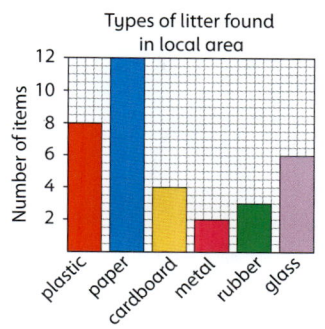

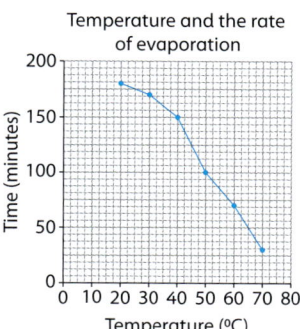

A bar chart is used when there are separate categories or types of things being studied. These are on the horizontal axis as separate bars.

A line graph is used to plot individual points where the values on the horizontal axis and vertical axis are both numbers. The points are then joined together to make a continuous line.

A scatter graph is like a line graph but the points do not show a simple relationship. Instead, they are not joined up but still show a pattern or trend.

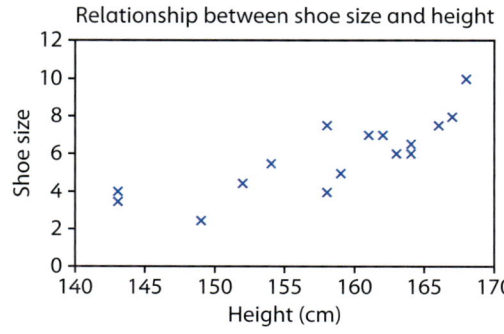

The shoe size of the people is plotted on the y-axis. The heights of the people are plotted on the x-axis.

> Why is the scatter graph better here than a bar graph or line graph? Do all of the people of the same height have the same shoe size? Explain your answer.

Drawings, photographs and videos

You have worked with scientific drawings before. Remember they are not like the pictures you paint. Scientific drawings are much simpler.

Scientists also use modern technology to take photographs and video clips of their investigations and results.

Photographs show a lot of detail

This is a very accurate way to record results. This level of detail would not be possible without using a camera.

Filming allows us to see things that may be impossible to see in person. Scientists can observe what happens to shadows during the day and time the details accurately by slowing down a film and piecing it together. This is called time-lapsed filming.

Research on the internet time-lapse films of shadows or the Sun seeming to move across the sky. Choose the best one to share with your class. What does the film show that you could not see with drawings or photos?

Drawing conclusions

The last stage of an investigation is when scientists look at their results carefully. It is at this stage that they make sense of their results. They work out if the results have helped them to answer their investigation question.

The questions they might ask are:

Can I *see* any patterns?

Is there a causal relationship in the data: did one thing cause another thing to happen?

Are any results unusual? Should I repeat any parts of the investigation?

Was my prediction correct? Does the evidence support my ideas?

How much do I trust the results?

Do secondary sources of information support my ideas?

Are further tests needed?

Scientists also link their conclusions to bigger scientific ideas. For example, if they are thinking about circuits and voltage, they will link this to their knowledge of electricity and components. They will also think about other factors, such as how many components are in a circuit and how they are arranged. They will even think about wider examples such as the applications of circuits and safety with electricity. They may even consider inventions such as electronic devices, including laptops, smartphones and digital cameras.

After completing an investigation, a good scientist will study their results and think about what went well and what could be improved. This is called evaluation and is an important part of the investigation process.

Presenting ideas

Scientists present their ideas by talking to others informally or at more formal meetings and conferences. They also write reports or make displays. This might be in a poster or computer presentation. They may include models.

Scientists are very careful to use the correct scientific language. This makes their ideas much clearer. They use standard units so their findings make sense across the world.

They also plan their reports and presentations to match the audience. For example, if they are talking to people who are not scientists, they will not include as much detail as they would in a more formal scientific paper.

Tips for presenting ideas

- Plan on paper first.
- Discuss your work with your team and share out the jobs.
- Think about your audience.
- Do not put too much information on a slide, poster or web page.
- Make any text, pictures and models eye-catching and clear.
- Use headings, colour and lists.
- Clearly set out what you did and what you found out.
- Show how your work leads onto further work.
- Use secondary sources of information and give credit to the people whose work you are using.
- Practise your presentation.
- Enjoy sharing ideas.

It is useful to fill out an investigation planning form. This sets out all the stages of your investigation. It helps you to remember everything you need to think about. Your teacher may give you one of these.

1 Classification and Habitats

In this unit you will:

- group living things using classification systems
- explore the reasons for classifying living things based on their characteristics
- find out how humans have positive (good) and negative (bad) effects on the environment
- learn about a number of ways of caring for the environment.

What different habitats can you see in this photograph?

Are these habitats healthy or damaged?

characteristic   classification
conservation   deforestation
environment   greenhouse effect
habitat   key   kingdom
microorganism   pollution
species

This is an insect. Discuss how you would recognise it as an insect and not as a bird or a mammal.

Think of some characteristics that help you to identify animals.

What has happened to the trees in the photograph? Why? How might this be a problem for insects and other animals in the area?

Science fact

Coal, oil and natural gas form from the remains of plants and animals that died millions of years ago. That is why they are called fossil fuels.

15

■ For more activities, go to Workbook 6 pages 14–15.

Classification systems

In this lesson you will learn how living things are classified into broad groups.

Think back

What characteristics are used to divide animals with backbones (vertebrates) into smaller groups or classes? List the classes.

Key words

characteristic
class
classification
group
kingdom
microorganism
species

No one can learn about or remember all of these different living things. To help, scientists do two things:

- they divide the living things into smaller groups
- they have a system for naming living things.

Classification

Classification is the grouping of living things based on characteristics they have in common. The characteristics are often physical features.

The main groups of living things are shown below. These are known as kingdoms.

Study the animals in this photograph. How are they the same? How are they different? Which animal group would you place them into? Why?

Science fact

Scientists now estimate that there are nearly 9 000 000 species of living things on Earth.

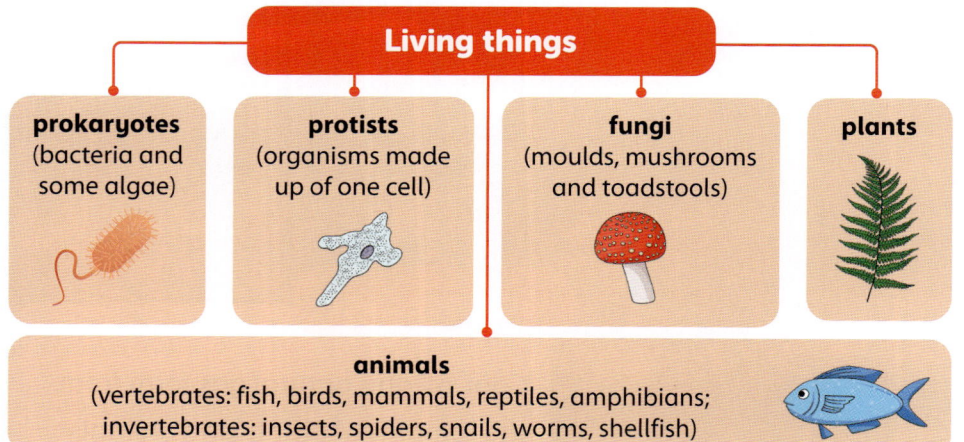

Living things

| prokaryotes (bacteria and some algae) | protists (organisms made up of one cell) | fungi (moulds, mushrooms and toadstools) | plants |

animals
(vertebrates: fish, birds, mammals, reptiles, amphibians; invertebrates: insects, spiders, snails, worms, shellfish)

Prokaryotes are also known as microorganisms. These are so small they can only be seen using a microscope. Most are very useful and help to recycle things in nature. Some are harmful and cause diseases.

■ For more activities, go to Workbook 6 page 16.

Plants can be divided into smaller groups too.

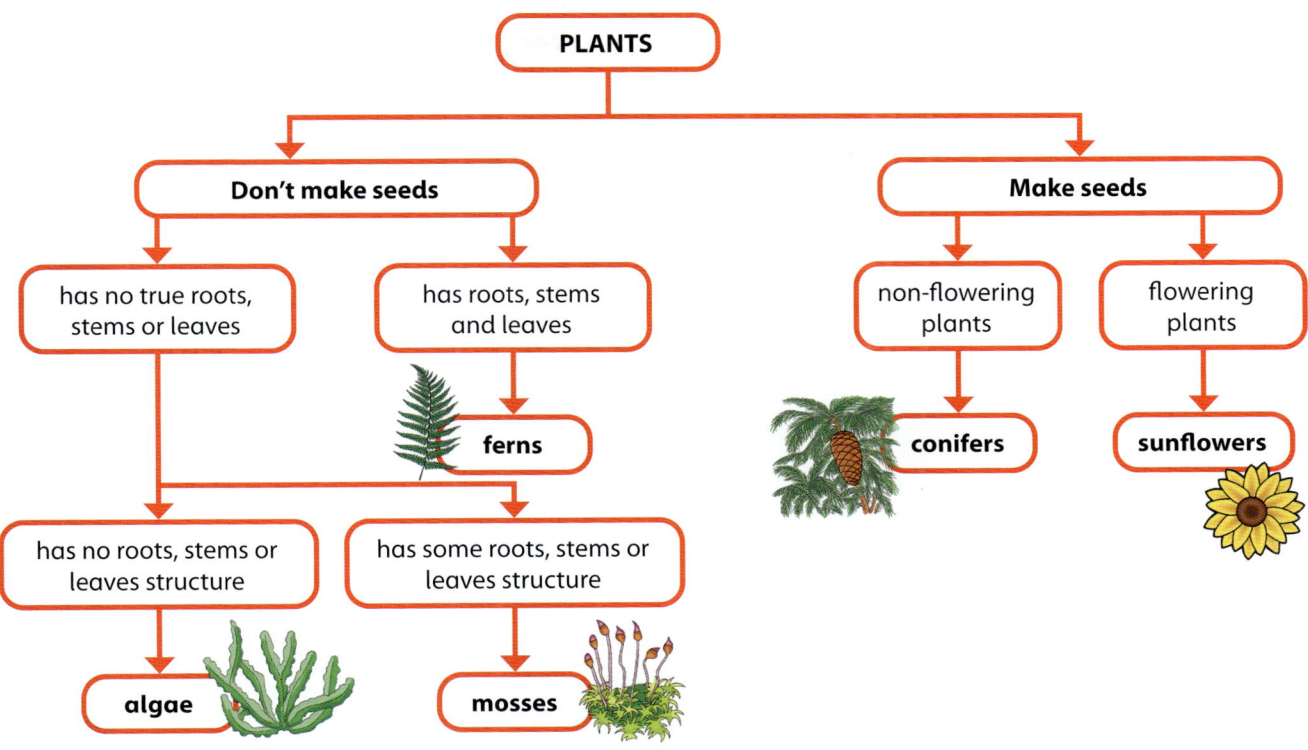

Plant classification survey

1 Use the diagram above to help you find some plants from each major group in your area.

2 Record the types of plants you find.

3 Present your findings as an information leaflet.

A naming system

In 1758, a Swedish scientist called Carl Linnaeus created a naming system for living things. We still use this today.

Living things are given a first name that puts them into a large group called a genus. The second name puts them into a smaller group called a species.

Using the Linnaeus naming system, the peregrine falcon is called *Falco peregrinus*. It is written in italics to show it is a scientific name.

Key idea

Living things are classified into smaller groups using characteristics they have in common.

Stretch zone

Find out the scientific names for a great white shark and a merlin. Which one is related to the peregrine falcon? Explain why.

Using characteristics to classify animals

In this lesson you will explore how animals are classified according to their characteristics.

Key words

characteristic
classification
habitat

Think back

Think back to what you have learned about classification.

What are the main groups of animals?

Polar bears are adapted to live in cold climates. They have thick fur to help them to stay warm. Their feet are adapted so that they can grip and walk on icy surfaces. They can run away from predators because they have strong, muscular legs. These are characteristics of a polar bear.

What kind of habitat do these animals live in?
Name three other animals you would include in the same vertebrate class.

Polar bears have white fur. How does this help them to survive in the habitat they live in?

Classifying animals

1. Observe each of the animals in the photographs. List the main characteristics of each animal.

2. How does the animal use the characteristics to survive?

3. Use the characteristics to divide the animals into five smaller groups.

4. Which of your five groups contains more than one animal?

5. Which vertebrate class does not have an animal shown in the photographs?

■ For more activities, go to Workbook 6 page 18.

Galapagos finches

Scientists have studied the way that finches survived in the Galapagos Islands.

Charles Darwin classified the finches according to the shape of their beaks. If you look closely at the drawings, you can see why he thought their beaks were adapted to suit the different foods available where the birds lived.

Some examples of Galapagos finches

Does the shape of the beak affect the food picked up?

You will need a variety of seeds and a set of different types of plastic forceps.

Picking up seeds

Use a variety of seeds and one type of forceps for this investigation.

1 Place the seeds in a petri dish or on a clean sheet of paper.
2 Use one type of forceps to pick up the seeds. Make separate piles of seeds according to their characteristics.
3 Draw a table with a diagram of each type of seed, its measurement and the number of seeds you picked up.
4 Were some seeds easier to pick up than others?

Investigating different beaks

Use one type of seed and different types of forceps for this investigation.

1 Investigate how many seeds you can pick up and form a pile within 30 seconds with each forceps.
2 Test each forceps.
3 Does the shape of the forceps change the number of seeds you can pick up?
4 Write a conclusion to explain if you agree or disagree with Darwin's ideas about the link between finches' beaks and the food they eat.

Be a scientist

Scientific diagrams are different from drawings. They should include the important features with accurate labels and measurements.

▶ page 12

Warning! Do not eat any of the seeds. Wash your hands after the investigation.

Key idea

Animals are classified into groups by observing their similar and different characteristics.

Stretch zone

Anteaters feed on ants. Research how the anteater is adapted to eat this food. Draw a picture and label its adaptations.

■ For more activities, go to Workbook 6 page 19.

Using characteristics to classify plants and microorganisms

In this lesson you will classify plants and microorganism based on their characteristics.

Key words
characteristic
classification
microorganism

Think back

Think back to the different plant groups. Write them down.

Classifying plants

It can be difficult to classify plants because they have lots of different characteristics.

It is common to use the characteristics of flowers to classify flowering plants.

Discuss the characteristics of these two plants.
What are the main differences? What are the similarities?

Here are three flowering plants. How can you classify them further?

Science fact

Scientists group plants according to their toxicity. Lily of the valley is beautiful but is classified as number 1 on the toxicity scale because it can be fatal if eaten.

Classifying local plants

You are going to survey and classify plants in your area.

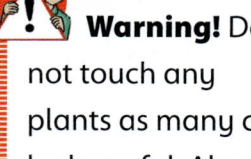
Warning! Do not touch any plants as many can be harmful. Always wash your hands after investigations.

1 Observe the plants on your survey. Design a table suitable for recording the characteristics of the plants you find.

2 Classify the plants as flowering plants or non-flowering plants.

3 Draw a diagram of the characteristics of the plants.

4 Describe how the characteristics might help the plant to survive.

■ For more activities, go to Workbook 6 page 20.

Classifying microorganisms

Microorganisms are living things that can be classified according to their characteristics. There are seven main types of microorganism but we will observe the five basic types. These are viruses, bacteria, fungi, protozoa and algae.

Science fact

Viruses can be classified by their shapes. Coronavirus is named because under a strong microscope it appears as a sphere with a crown or 'corona' of spikes on its surface.

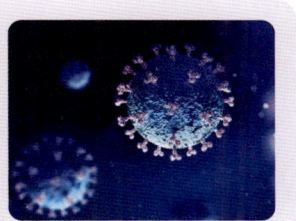

Aside from shape and size, what other ways could microorganisms be classified? Discuss with your partner.

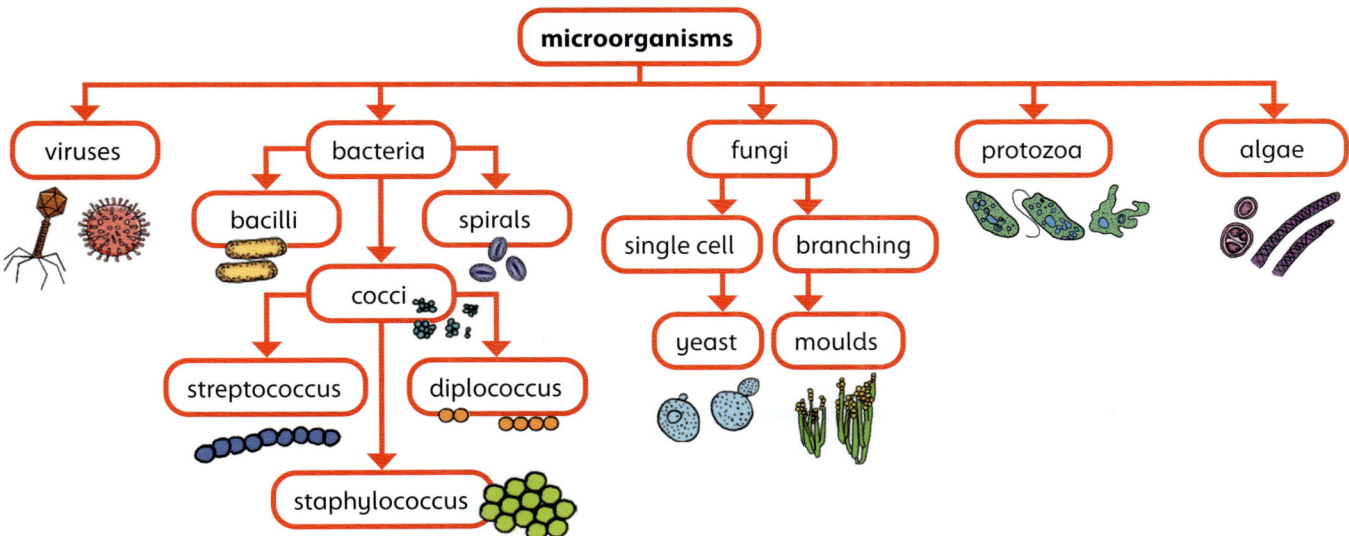

As you can see from the key, bacteria can be classified into three main types:

- bacilli – rods
- cocci – spheres like tiny balls
- spirals.

The cocci can be classified even further into streptococcus (chains), staphylococcus (clusters) and diplococcus (pairs).

Scientists use the shapes of bacteria to classify and identify them.

Researching microorganisms

Research the five types of microorganism above.

1 For each one, name an example and describe a disease it can cause.

2 Make a short information leaflet about your findings.

Key idea

Plants and microorganisms can be classified and grouped based on their characteristics.

■ For more activities, go to Workbook 6 page 21.

Using classification keys to group living things

In this lesson you will use classification keys to group plants and animals.

Key words

characteristic

classification key

Think back

What is a classification key?

Why do scientists need to use keys? What other techniques do they use to make sure they get an accurate count in an area?

Be a scientist

Scientists use keys to help them to count and classify marine plants, corals and fish. They can then measure which are surviving well and which might be in danger.

▶ page 10

In any survey of living things, classification keys are very important. For example, when scientists study invertebrates they need to use a key, as there are millions of different types.

Classifying invertebrates using a key

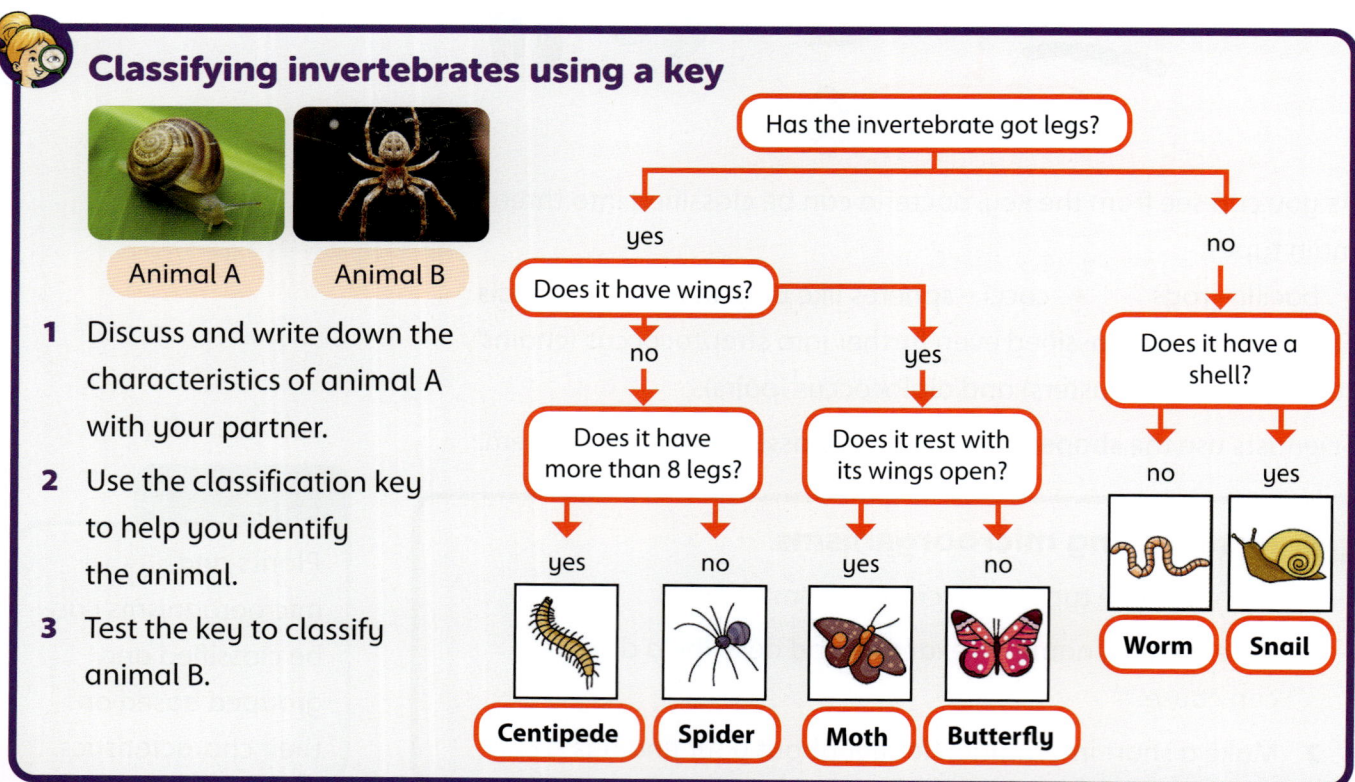

Animal A   Animal B

1 Discuss and write down the characteristics of animal A with your partner.

2 Use the classification key to help you identify the animal.

3 Test the key to classify animal B.

Has the invertebrate got legs?

yes → Does it have wings?

no → Does it have a shell?

Does it have wings? — no → Does it have more than 8 legs? — yes → Centipede / no → Spider

yes → Does it rest with its wings open? — yes → Moth / no → Butterfly

Does it have a shell? — no → Worm / yes → Snail

■ For more activities, go to Workbook 6 page 22.

Which class is it?

1 Use the classification key below to group this animal. Scientists sometimes have to use what they already know about living things to use a key. They will not rely only on what they can see when they look at one individual.

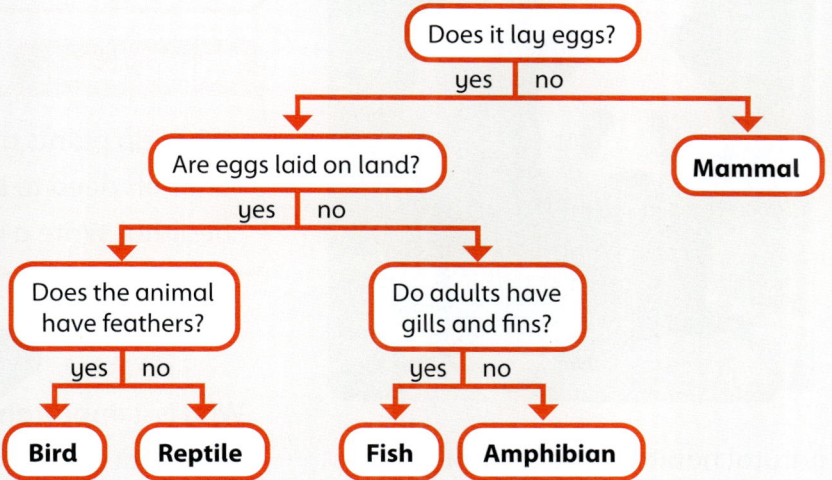

Does it lay eggs?
yes / no

Are eggs laid on land?
yes / no

Mammal

Does the animal have feathers?
yes / no

Do adults have gills and fins?
yes / no

Bird | Reptile

Fish | Amphibian

2 Did the key work?

3 Did you classify the animal in the photograph correctly? Write down the class it belongs to.

Designing a classification key

Think about a living thing that you have seen.

You are going to design a key to classify the living thing.

1 Start by listing its characteristics.

2 Introduce each characteristic with a question that has a 'yes' or 'no' answer. Start with the most important characteristic.

3 Write down your key – use keys you have already seen to give you some clues.

4 Test your key on other students.

5 Evaluate your key. Do you need to change anything?

Key idea

Living things can be classified and identified using keys. This is important as it would be impossible for anyone to remember every living thing.

Stretch zone

Design a key that can be used to divide the kingdom of plants into flowering plants and non-flowering plants.

■ For more activities, go to Workbook 6 page 23.

Looking after our world

In this lesson you will find out how humans have positive and negative effects on the environment.

Key words
conservation
endangered species
environment
extinct

Protecting the environment

There are millions of different species of living things on Earth. To be in the same species living things must be very similar and be able to have offspring together. Some species are in danger of dying out and becoming extinct. We call these endangered species.

A habitat is where an animal or a plant lives. Hunting and cutting down forests can destroy natural habitats and the animals and plants that live there. There are many plants we have not yet discovered. They may help us with medicines or other useful things.

Think back

What do plants and animals need to be healthy? Write a list.

Why is it important to protect endangered species and habitats?

What can we do to protect our environment?

Throughout the world caring for the environment is big news. There are many organisations that work to help protect the environment. One of these organisations is the World Wide Fund for Nature (WWF). It works to protect the future of the natural world. The WWF started its conservation work in 1961.

We can do lots of things to help protect our environment. Here are some ideas:

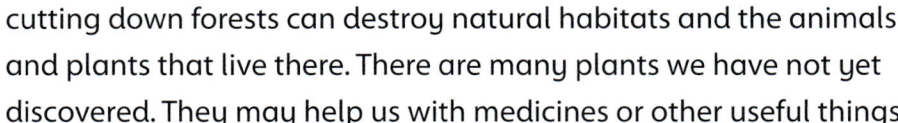

Teach others how important it is to protect the environment.

Replace trees in forests. Repair damaged habitats.

Encourage others to care for the environment.

Establish parks and nature reserves.

Show you care by doing simple things. Here is something you can do to help.
Remember to turn off lights when you leave a room.

Remember TREES!

Investigate a local habitat

You are going to investigate a habitat in your local area.

1 Try to identify the plants and animals that live in the habitat. The pictures show you some clues you can look for. Write down what you see on the identification key your teacher gives you.

webs   trails
chewed leaves
eggs

digging
footprints   fur   hair
damage to trees and leaves
scratching

birdsong
feathers   eggshells
eggs   nests

What do they look like?
Look for cacti, trees,
small plants and so on.

skin   movement

2 Create an information leaflet about the habitat. Include tables, bar charts or line graphs of your findings in the leaflet. Identify what can be done to protect the habitat for the future.

Science fact

Animals and plants are adapted to their habitats. If the habitat changes, they may not be able to survive. They will have to move or die.

How is the change in this habitat affecting the polar bear?

Stretch zone

Imagine you have been asked to create a small natural habitat near your school. Design a habitat for amphibians and fish. List three things the habitats must have to keep the animals healthy.

Key idea

We can affect our environment in positive and negative ways.

■ For more activities, go to Workbook 6 page 25.

Air pollution

In this lesson you will learn about how air pollution has a negative effect on the environment.

Key words

acid rain
atmosphere
climate change
fossil fuel
greenhouse effect
pollution

Think back

What is the definition of pollution? List three examples you know about.

Look at the photograph. What is being sent out into the atmosphere? How could this be reduced?

Acid rain

Humans add many materials to the environment. If these are harmful, they are known as pollution. Pollution that enters the atmosphere from factories, cars, buses and lorries, and homes makes rain more acidic. This rain is known as acid rain.

When fossil fuels such as coal and oil are burned in power stations, factories or in our homes, they make acidic gases. Most of these acidic gases are mixed into the air. When the gases mix with the clouds it can cause rain to become more acidic. Acid rain can damage living and non-living things.

Look at the photographs of the building and the trees. How have the trees and building changed because of acid rain?

List some ways that acid rain could be reduced.

■ For more activities, go to Workbook 6 page 26.

The greenhouse effect

The Earth is surrounded by the atmosphere. The atmosphere is made up of a mixture of gases which trap some of the Sun's heat energy. This warms the Earth so that it can support life. This is called the greenhouse effect.

Scientists think that human activities increase the amount of gases in the atmosphere. This can lead to too much heat being trapped. This is known as the enhanced greenhouse effect. It is one of the reasons why people are concerned about climate change. The gases the scientists are worried about are:

- carbon dioxide – this comes from burning fossil fuels and cutting down trees
- methane – this comes from rice paddies and from animals such as cows.

How do you think we can reduce the greenhouse effect?

Science fact

Climate change is when the weather patterns in a particular area change over a long period of time.

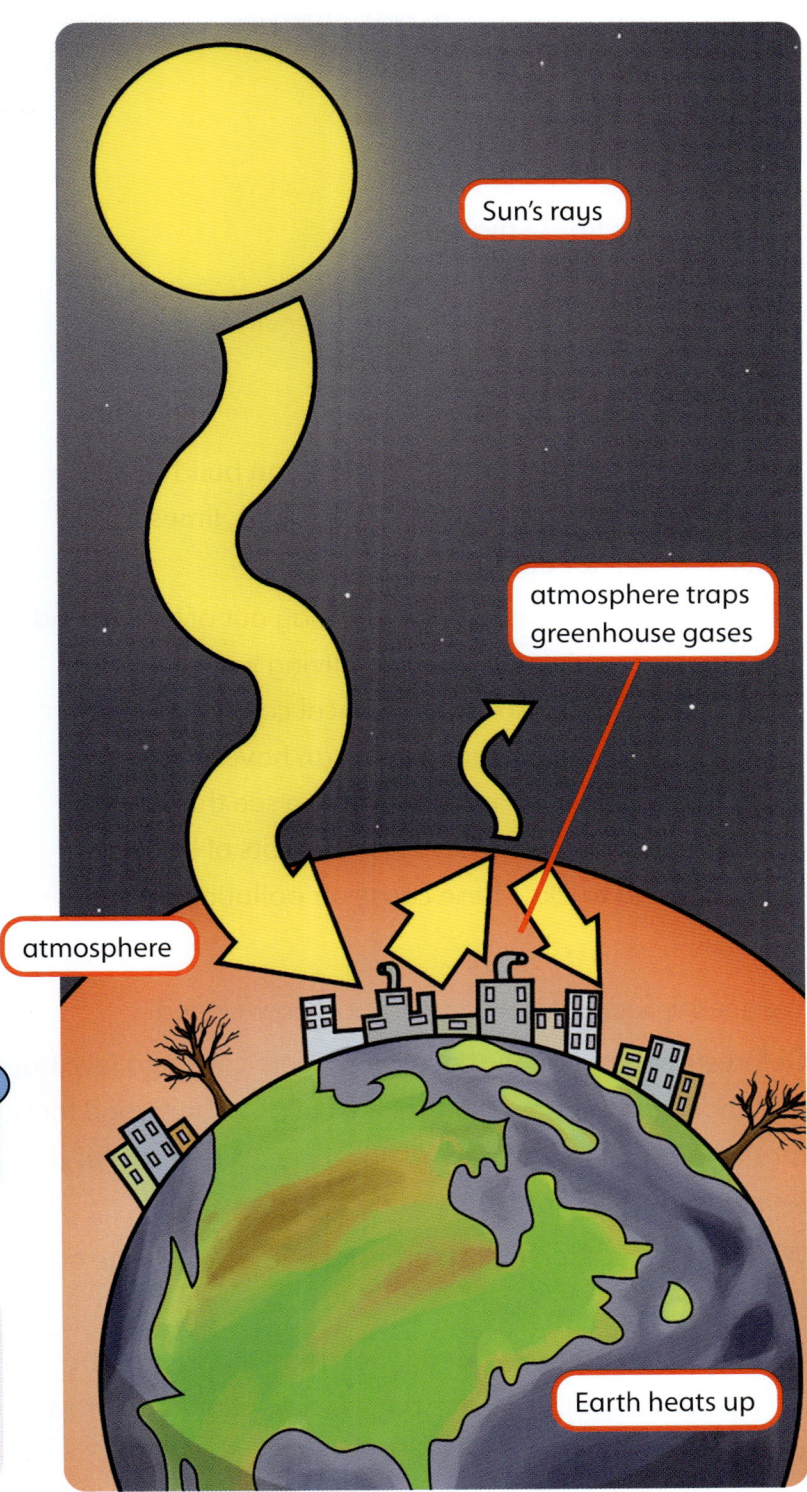

Sun's rays

atmosphere traps greenhouse gases

atmosphere

Earth heats up

Stretch zone

Research how the greenhouse effect is causing ice to melt and sea levels to rise. What is this called? Produce a report of your findings.

Key idea

Burning fossil fuels produces gases that pollute the air.

■ For more activities, go to Workbook 6 page 27.

Digging up and cutting down habitats

In this lesson you will learn how quarrying and deforestation have negative effects on the environment.

Key words
deforestation
erosion
quarrying
variable

What is your house made of?

Where did these materials come from?

How could this have damaged the environment?

Quarrying

We need certain materials so that we can build homes, shops, factories and offices. We need stone, limestone, chalk and clay for our buildings.

We get stone, limestone, chalk and clay out of the ground by quarrying. Local people earn a living working in the quarry. Quarrying also gives the local community an income. Unfortunately, quarrying can have a negative impact upon the environment. You can see this damage in the photograph. Quarries also need lots of heavy machines and trucks. These cause air pollution.

A quarry site

What do you think quarrying has done to this local environment?

Deforestation

We need wood. But how much wood do we need?

Deforestation means cutting down trees. Not just a few, but hundreds and thousands of trees. Forests are cut down for wood for buildings, furniture, doors and paper making. Forests are also cut down to make room for farming and for building new houses.

Why is deforestation a problem? If we cut down large areas of trees many things can happen.

- When large areas are cleared of trees, the soil can be washed away. This means that no new plants can grow.

- Trees absorb rainfall. If deforestation happens in mountainous regions, there are no trees to soak up the rainfall. This can cause flooding in the lower regions.

- Trees can help produce rainfall. Without forests the climate may become drier.

Trees have been cleared in this area

28

■ For more activities, go to Workbook 6 page 28.

How do trees reduce soil erosion?

You are going to investigate soil erosion.

1 Set up three bottles as shown in the diagram. Make identical slopes of soil in all three bottles.

2 In the first bottle, add small twigs and branches. In the second bottle, grow grass seeds or push twigs or sticks into the soil. Leave the other bottle as bare soil.

3 Gently pour water from a watering can onto your slope. Use the same volume of water for the same amount of time. Catch the water in a small cup or bottle.

4 Observe what happens.

5 Write a report of your investigation. Describe the independent, dependent and control variables.

Be a scientist

Scientists alter only one variable so the rest, such as angle of the slope, will be kept the same.

▶ page 9

How can we reduce the effects of deforestation? We can plant trees to replace the ones chopped down. We can stop chopping down very old forests.

Quarrying and deforestation are two human activities that have a negative effect upon the environment. But there are many more, and you will explore some in the next lesson.

Stretch zone

Plan a local survey to identify some of the ways in which slopes are prevented from washing away in the rain. Design an information leaflet to share your findings.

Science fact

Scientists have calculated that half of the Earth's forests have already been cleared. At this rate, all of the forests will disappear in about 100 years.

The world's last tree

How do trees help to prevent soil erosion?

How else can slopes be protected from erosion (wearing away)?

Key idea

Getting the raw materials we need, such as stone and wood, can cause damage to the environment.

Water pollution and waste disposal

In this lesson you will explore human activities that can lead to the pollution of water.

Key words

acid

alkali

landfill

pH scale

pollution

sample

waste

Pollution is when humans introduce materials that harm the environment. Human activities that can pollute water include:

- throwing away waste materials such as plastics
- oil from boats
- chemicals from factories and homes
- pesticides and fertilisers from farms
- sewage.

The pollution can enter food chains and food webs and could kill animals and plants. It may also spread human diseases.

These water samples were taken from the same river. One was taken before it flowed through a town and the other after.

Why is one sample a different colour to the other one?

Would you risk drinking either of the water samples?

Look at the photograph of the river. How safe would the water be to drink or bathe in?

What could be done to improve this river habitat?

Science fact

Up to 70% of waste from industry is dumped into water. Sewage from houses causes up to 80% of water pollution.

30

■ For more activities, go to Workbook 6 page 30.

Scientists can test to see if water is acidic or alkaline. They do this by using indicators. These change colour in acids and alkalis. One example is the universal indicator. Each colour is given a number. This is the pH number. Pure water has a pH number of 7. It is neither acid nor alkali – it is neutral.

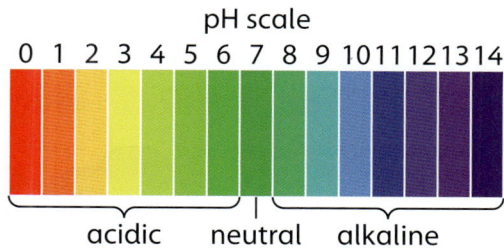

pH scale

0 1 2 3 4 5 6 7 8 9 10 11 12 13 14

acidic   neutral   alkaline

Testing water samples

You will be given five water samples to test.

1 Test each sample with an indicator. Check the colour against the pH scale.

2 Record the pH number for each sample.

3 Which samples were acidic? Which samples were alkaline? Were any samples neutral?

Warning! DO NOT drink water from an unknown source. Even if a water sample is neutral, it does not mean it is pure. It could still have harmful microorganisms or chemicals in it.

Stretch zone

Design a test to find out how cloudy or clear water samples are. Test your method on the samples your teacher gives you.

Waste disposal

Discuss some of the problems caused by disposing of waste. How is waste treated and disposed of in your area?

Waste produced by homes, schools, hospitals and factories has to be disposed of. One method is to bury the waste in landfill sites. This can cause problems by attracting rats and flies. Some chemicals can wash into streams and rivers.

In some places the waste is burned to provide energy.

Key idea

Some human activities, such as disposal of farm and factory chemicals, sewage and plastics, can lead to the pollution of water.

■ For more activities, go to Workbook 6 page 31.

Caring for the environment

In this lesson you will explore ways to save electric energy or use alternative energy sources.

Key words
energy consumption
non-renewable energy
renewable energy
solar panel

Think about a typical day in your life. How is your food cooked? Where do you go? What energy do you use?

Power stations produce the electricity we use every day

Our environment needs our help to protect habitats and endangered species. We need to think about the local and the global (worldwide) environments. Many of our activities affect the local and global environments. Here are some ways to help care for the environment.

Using energy

We use electrical energy for lots of things. This is called energy consumption. If we want to care for the environment, we need to think about the energy we use. We also need to think about how our energy is produced.

Saving energy

One of the things we can do to care for the environment is to use less energy in our everyday lives. When you walk out of a room at night, do you turn off the light?

Producing energy can cost a lot of money. We pay for the energy we use. So, if we save energy, we can save money and also help to care for the environment.

What types of things use energy? List some examples.

Science fact

Scientists have calculated that 75% of the electricity used to power devices is used while the products are 'on standby'.

32

■ For more activities, go to Workbook 6 page 32.

School energy survey

You are going to survey your school to record the different uses of electricity.

1 Make a note of any electrical devices you see.

2 For each one, write down an example of how the device could be used less.

3 Produce an energy-saving leaflet for your school based on your findings.

Describe how energy use could be reduced and why this would help the environment.

Be a scientist

Scientists design and construct results tables before starting an investigation or survey. This helps them to plan and ensures they collect data in an organised way.

▶ page 11

Renewable and non-renewable energy sources

Using fossil fuels such as coal and oil cause pollution. These fuels are being used up and will run out. They are non-renewable.

There are different ways to produce the energy we need. Solar (Sun) power is used in solar panels. These capture sunlight energy, which is converted into electrical energy for use in the home. The Sun will not

run out and so this is a renewable source of energy. Other renewable sources of energy are hydroelectric power (from water), geothermal power (from the heat inside the Earth) and wind power.

Renewable energy sources do not produce as much pollution as non-renewable energy sources. We call this environmentally friendly energy production.

Discuss the advantages of renewable energy.

List any examples you have seen in your local region.

Stretch zone

Research energy-saving bulbs.

Explain to your partner how much electricity could be saved in your school over a year by changing all the lights to this type of bulb.

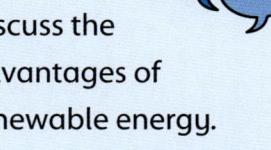

Key idea

Saving electrical energy can reduce the pollution from burning fuels.

■ For more activities, go to Workbook 6 page 33.

Recycling and reusing materials

In this lesson you will learn how recycling and reusing materials can help us to care for the environment.

Key words
recycle
reduce
reuse

Do you like to drink bottled water or fizzy drinks? What is the bottle made of?

When you finish your drink, what do you do with the bottle?

We can help our local and global environments in many ways. One way is to recycle. Recycling is a way of reusing things we no longer need. Recycling also helps to reduce waste.

One of the things we can recycle is glass. We use glass for many things such as drinks bottles and food jars. If we recycle our empty bottles and jars we can make new glass products.

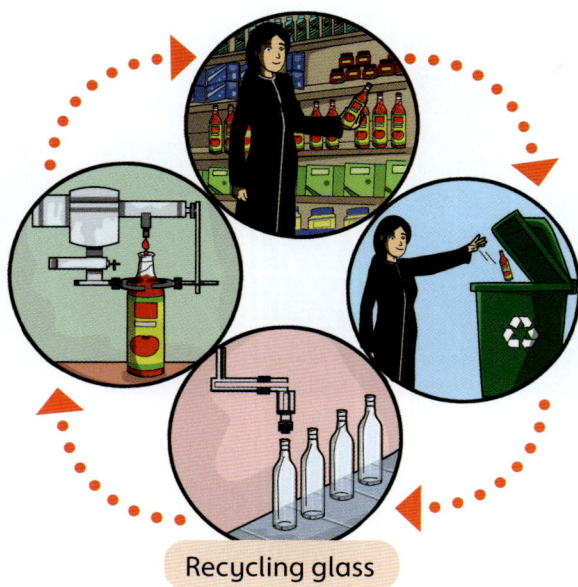

Recycling glass

We use paper in lots of things such as newspapers and paper bags. We can also recycle paper. Then we can save trees!

Discuss with your partner what other things can be recycled.

Recycling paper

34

■ For more activities, go to Workbook 6 page 34.

Recycling action plan

You are going to work in a team.

1   Design an action plan to encourage people in your classroom to recycle.

- Start by finding out how much is thrown away each day.
- Work out which of these things can be recycled.
- Plan how these can be collected and stored.

2   Share your plan with the class. Evaluate the advantages and disadvantages of each plan.

3   Agree a class plan. Include:

- what you are hoping to achieve – your targets
- what you will need
- who is responsible for what
- how you will know if you are successful.

4   Make a poster version of your plan to display in the classroom.

Look at this picture of a classroom and compare it to your own classroom. Discuss what is the same and what is different. Identify some of the things that can be recycled.

Science fact

Scientists have calculated that recycling just one aluminium can can save enough electricity to power a TV for three hours.

 Stretch zone

Find out about the system used to label plastics so that we know if they can be recycled or not. Take a photograph of an example label on food packaging or a bottle and explain the information to your class.

Key idea

Recycling and reusing materials can help to reduce pollution and energy use.

■ For more activities, go to Workbook 6 page 35.

Managing litter

In this lesson you will investigate how managing litter can help us to care for the environment.

Key words

landfill

litter

What do you do with your litter? Do you put it in the litter bin? Do you throw it on the ground?

What sort of things do we think of as litter? Think about the food and drink you have. Think about the sort of packaging it comes in.

What is litter?

Litter is any waste material, for example empty drinks bottles, that people throw away carelessly. Litter can be found in many places.

Often when people drop litter they think, 'My small piece of litter will not make a difference.' Imagine if we all thought like that. Then we would have big problems!

Where can we find litter? Write down three places where we can find litter.

Litter survey

You will adopt an area near to your school. Your class will be responsible for planning ways of improving this area by reducing litter.

Draw a plan of the area. Imagine you are flying above it and draw what you would see.

1. Carry out a survey to find out how much litter is in the area.

2. Identify and count different types of litter, for example:
   - paper and cardboard
   - metal
   - plastic
   - food waste
   - glass
   - wood
   - polystyrene foam
   - other

3. Design a results table to record your findings.

4. Present the findings in a suitable chart or graph. Think about which type of chart is the best for the data you have collected.

5. Now discuss how you can reduce the litter in the area. Write up your plan. Include where you might place litter bins and information posters. Try to carry out your agreed actions.

6. Carry out a second survey one month after the first one. What impact did your actions have on the amount of litter?

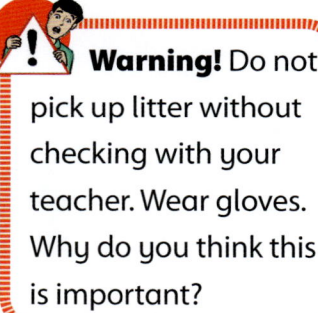

Warning! Do not pick up litter without checking with your teacher. Wear gloves. Why do you think this is important?

■ For more activities, go to Workbook 6 page 36.

Why do we need to care about littering?

Littering causes many problems. Litter is not very nice to look at. It is not nice to walk past litter on the way to school, home or work. The problem is more serious than this. Litter can harm people because it can attract rats that carry diseases. Some litter may contain broken glass that can harm people and animals. As litter builds up it needs to be put somewhere else so that it does not harm people or animals. By carelessly dropping litter we put ourselves and wildlife at risk.

Imagine a hungry animal trying to find food. It smells food but it isn't food. It is food packaging. There may be a small amount of food in the packaging. But usually there is a lot more packaging than food. The hungry animal wants to eat. It eats some of the packaging. Unfortunately, many sorts of packaging cannot be digested by the animal. Packaging can cause many health problems.

We know how important it is to not drop litter. In some places litter is a big problem.

The Maldives are very beautiful islands and are visited by lots of tourists. The tourists and the islanders produce lots of litter and general waste. To dispose of all this litter and waste, one of the islands is now used as a landfill site.

Science fact

A lot of litter ends up in the oceans. Scientists have calculated that 9 billion tonnes of waste enters the oceans every year.

Stretch zone

Find out what happens to all of the waste material in your area. Where is it taken to? How is it treated?

Whose job is it to stop or reduce littering?

Is littering each person's responsibility? Or is littering the responsibility of the community?

Discuss these questions in your group. Draw a spider diagram showing your discussions and present it to the class.

Did all the other groups have similar discussions?

Key idea

We can all care for the environment by tidying up and not dropping litter.

1 Classification and Habitats

37

■ For more activities, go to Workbook 6 page 37.

Protecting the environment

In this lesson you will plan some ways of working together to care for the environment.

Key words

acid rain
environment
litter

Think back

What can you do to help others to care about the environment?

In this lesson you will work on an investigation to encourage others to care for the environment. Below is some information about investigations you can carry out. Each investigation asks you some questions. You will need to do some research to find the answers.

At the end of the investigation you will give a presentation. Think about the sort of presentation you can give. Maybe you can perform a play. Maybe you can design a poster that all the school students can see.

Investigation 1: Making our community a tidy place to be

This investigation is about littering.

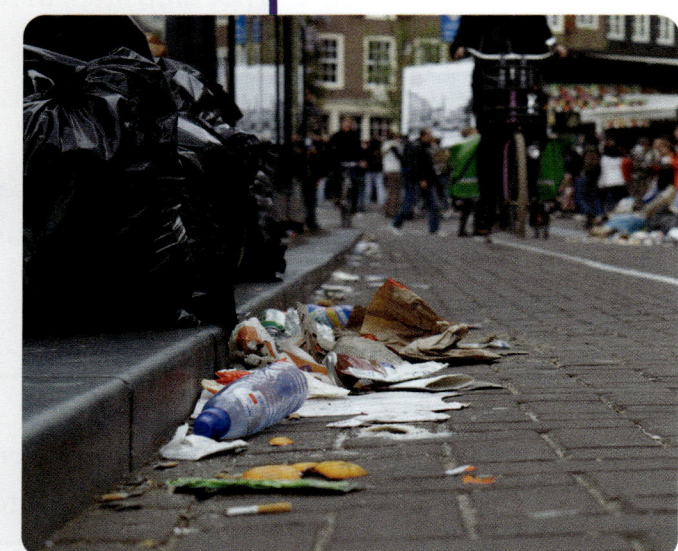

In your group, talk about some of these questions.

a   Is our school tidy or is there any litter?
b   Is there any litter on the street outside school?
c   Is there any litter near our homes?
d   Are the play areas clean and safe to play in?

How can you carry out this investigation? Here are some ideas.

- You can find photographs on the internet or in books of areas where litter is a problem.
- Who can you ask about litter and what happens to it?
- Where does the litter go when it is collected?
- How can we use this information to encourage people to care about the environment?

38

■ For more activities, go to Workbook 6 page 38.

Investigation 2: Saving energy

This investigation is about saving energy.

In your group, talk about some of these questions.

a In our school, do we waste energy or are we energy efficient?

b Does our school use renewable energy?

c Is there a factory or office that uses energy near our homes?

d How can we save energy in our homes?

How can you carry out this investigation? Use the questions in Investigation 1 to help you.

Investigation 3: The problems with acid rain

This investigation is about the problems with acid rain.

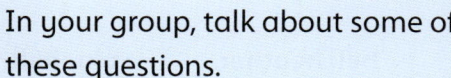

In your group, talk about some of these questions.

a What is acid rain?

b How is it caused?

c Why is acid rain a problem?

d What can we do about reducing acid rain?

How can you carry out this investigation? Use the questions in Investigation 1 to help you.

Key idea

We need to work together to look after our environment.

Check how much you know.
Try the questions on pages 40–41.

■ For more activities, go to Workbook 6 page 39.

1 Circle the correct phrase or word to finish each sentence.

a A vertebrate has   no backbone   a backbone

b The vertebrate classes are fish, amphibians, birds, mammals and   insects   reptiles

c Prokaryotes is a kingdom of living things that contains   bacteria   trees

2 Use the key below to identify the plants.

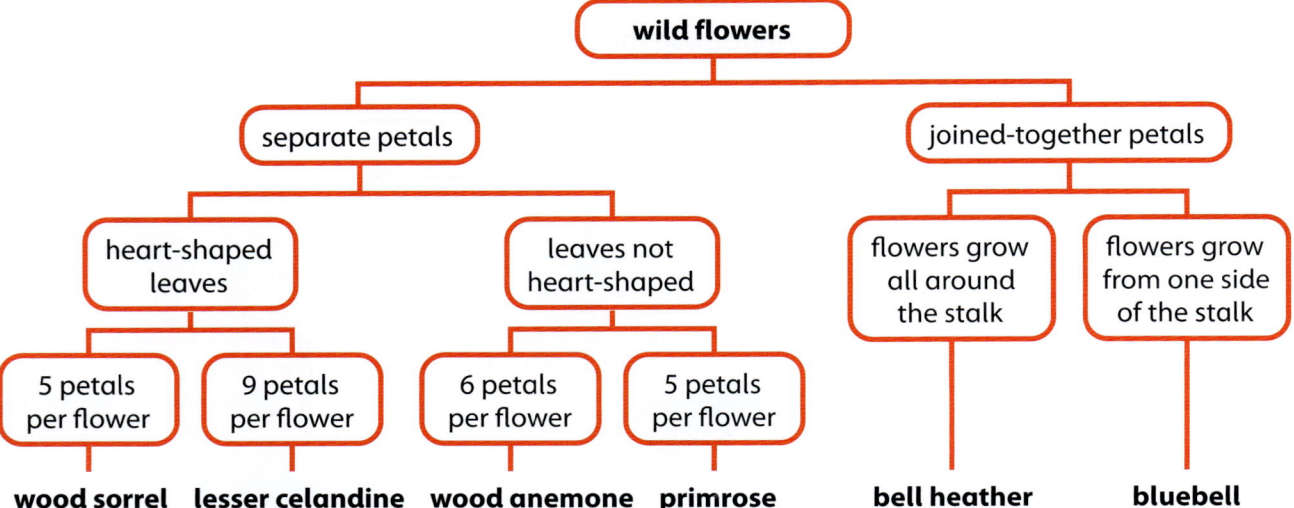

A: _____ B: _____

C: _____ D: _____

E: _____ F: _____

3 List two observable characteristics that you would use to classify:

a a plant

   1. _____

   2. _____

b an animal

   1. _____

   2. _____

■ For more activities, go to Workbook 6 page 40.

4 Fill in the missing words in these sentences.

Human activities can damage the e__v__r__ __ __ __ __t. Plants and animals living in their natural habitat may be end __ __ __ __ __d. If we don't protect them, they may become ex__ __ __ __ __.

5 Label the greenhouse effect diagram by drawing a line from each label below to the correct box on the diagram.

atmosphere   Earth heats up   Sun's rays

atmosphere traps greenhouse gases

6 Arrange the pictures about recycling glass bottles in the correct order. Put the number 1 in the box of the first stage and so on until stage 4.

  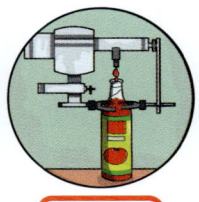 

7 Look at the table. It shows what has happened to the fish and coral found on coral reefs between 1990 and 2020.

	1990	2020
Number of fish species	527	390
Percentage of coral cover	66	12

a What happened to the number of fish species between 1990 and 2020?

_____

b What happened to the amount of coral on the coral reefs between 1990 and 2020?

_____

c Write down two possible reasons for the changes shown in the table.

_____

_____

8 Write down two ways to encourage your friends and family to care for the environment.

_____

_____

■ For more activities, go to Workbook 6 page 41.

In this unit you will:

- learn where the major organs are to be found in the human body
- find out about the functions of the major organs
- explore the human circulatory system and describe the functions of the heart, blood vessels and blood
- describe how nutrients and water are transported in humans and animals
- learn about infectious diseases and how we can protect others and ourselves from them
- recognise how diet, exercise, drugs and lifestyle affect our bodies.

Study the photograph. Which parts of the body are the runners having to use to travel so quickly?

Discuss why the runners have a higher heart rate after the race than before. What other changes might you notice about the runners after the race?

Science fact

The adult human heart beats between 30 000 000 and 40 000 000 times a year!

circulatory system
defence mechanism
digestive system   drug   function
infectious disease   lifestyle
medicine   nervous system
organ   urinary system
vaccine

What would happen to the runners if they had not eaten for three days before the race? Would they run faster or slower? Why?

■ For more activities, go to Workbook 6 pages 42–43.

Where are our major organs?

In this lesson you will learn where the major organs are to be found in the human body.

Key words

brain
heart
intestines
kidneys
liver
lungs
organs
stomach

Think back

How are the brain, heart and lungs protected?

We can think of the human body as three sections. They are the:

- head
- chest
- abdomen.

You can see these in the diagram of the body. The neck separates the head and chest. The waist separates the chest and abdomen. The body's major organs are inside the three sections of the body.

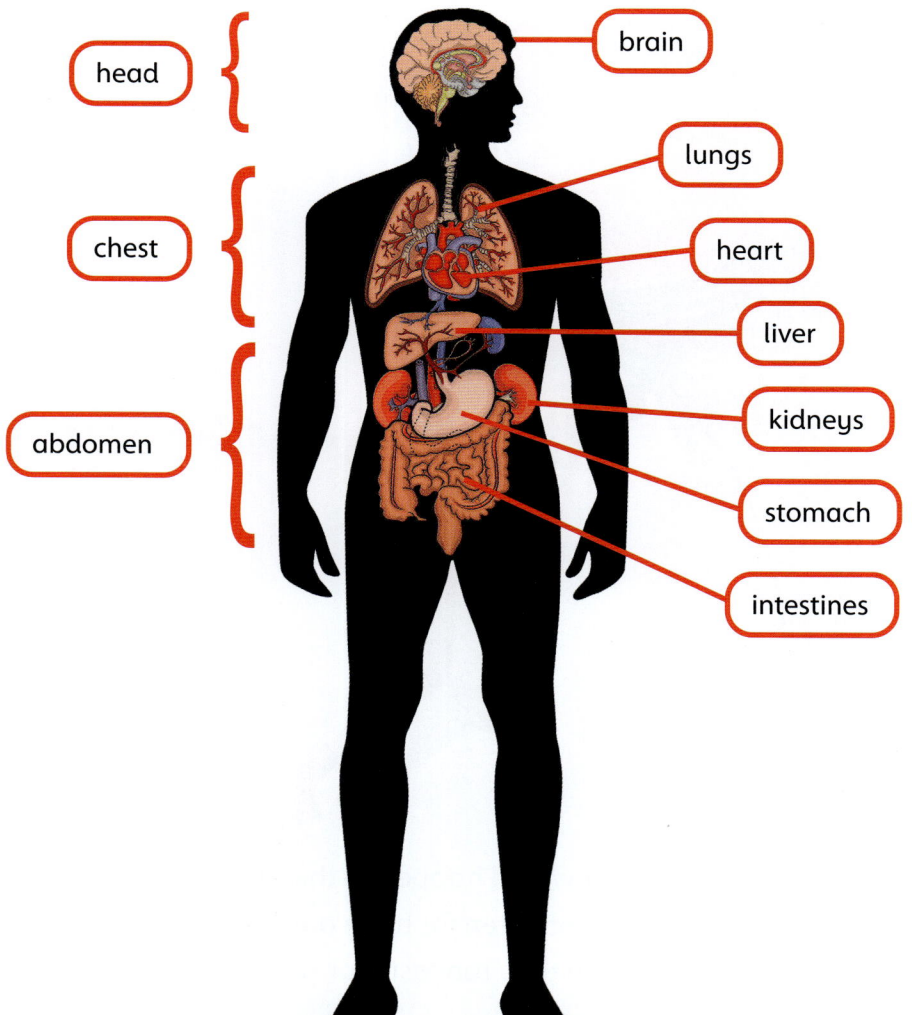

head

chest

abdomen

brain

lungs

heart

liver

kidneys

stomach

intestines

Look at the intestines in the diagram. Discuss why they have to be so long. Share your ideas with the class.

■ For more activities, go to Workbook 6 page 44.

Where do the organs fit?

In a pair or small group you are going to create a life-sized drawing of the position of human organs.

1 Place a large sheet of paper onto the floor. One person in your group should lie down on top of the paper.

2 Draw around the person carefully, so you make a body outline. The person can now get up. If you prefer you can cast a shadow onto the paper and draw around this.

3 Add drawings of the organs onto your body outline. Make sure they are the correct size and label each organ.

4 Display your body outline in your classroom and compare yours with other groups. Did you miss any organs?

Science fact

Many scientists think of the skin as an organ. It is made of layers and contains blood vessels, nerves, hairs and oil glands.

Organs are parts of the body that carry out important functions. They are made up of different cells and tissues. To carry out more complicated roles in the body some organs work together as a team. These teams are called systems.

 cell   tissue   organ   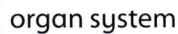 organ system    organism

  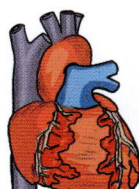 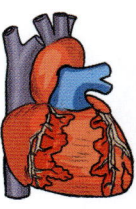 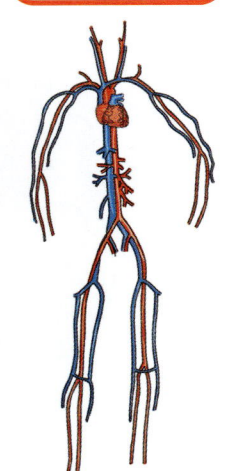 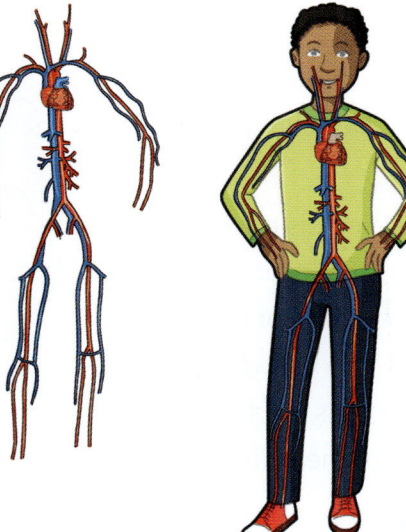

From a cell to an organism

Key idea

Each of the major organs in a human has a specific place in the body.

Stretch zone

Research three organs and write a paragraph about their function in the human body.

■ For more activities, go to Workbook 6 page 45.

What do our major organs do?

In this lesson you will find out about the main functions of the major organs.

Key words

breathing
circulation
digestion
excretion
function
system

Think back

How many of the major organs can you name? Can you remember where they are in the human body?

The human body is made up of different working parts. The parts that work together are called a system. Each major organ and its system has a special job to do. This is called their function.

Lungs

We breathe air into our lungs, which are part of the respiratory system. We cannot live more than a few minutes without air. Air contains oxygen gas which is vital for life processes that occur in the body. During these processes, the body produces a waste gas called carbon dioxide. Carbon dioxide poisons us if we do not get rid of it. It is removed from the body through the lungs.

The process of taking in air containing oxygen gas and pushing out air containing carbon dioxide is called breathing.

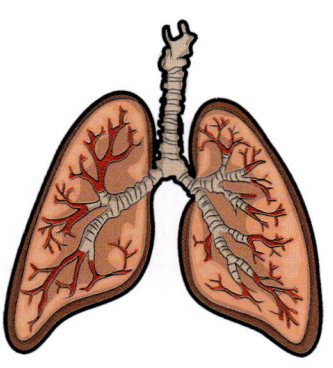

Discuss how the lungs allow gases to enter and leave the body. Why is this important?

Heart

The heart pumps blood around our body. Blood flows through the body in tubes called blood vessels.

The heart, blood and blood vessels make up the circulatory system. This circulatory system takes blood to and from every one of the millions of cells in our body. The blood carries oxygen and nutrients to the cells, and carbon dioxide and waste away from the cells.

Brain

The brain, spinal cord and nerves make up the nervous system. The brain is the control centre of the body. It does millions of tasks for you without you having to think about them.

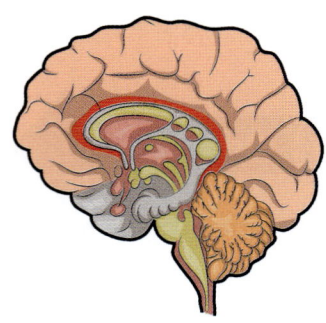

46

How quick are your reactions?

In this investigation, you will test to see how quickly you can move your muscles once your brain has detected movement.

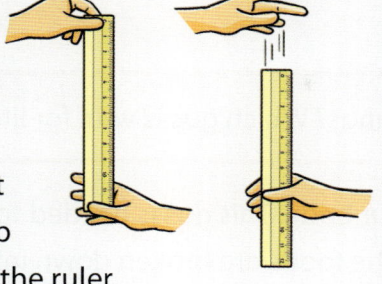

1 Ask someone to hold a ruler in front of you. Place your finger and thumb near the zero mark but don't touch the ruler.

2 The person holding the ruler should let go without telling you. When the ruler falls, try to catch it as quickly as you can.

3 See how many centimetres of the ruler have passed through your fingers before you caught it. Record this measurement. Do this three times to get an average distance.

4 The shorter the distance the ruler passed through your fingers, the quicker your reaction time. Compare the reaction times of the people in your group.

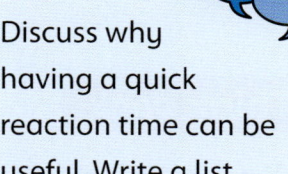

Discuss why having a quick reaction time can be useful. Write a list.

Do you predict your reaction times will vary at different times of the day? Why?

Stomach and intestines

Food goes in at one end of the body when we eat and waste comes out at the other end when we go to the toilet.

As the food we chew and swallow moves through the body it is broken down and absorbed into the blood. This is called digestion and takes place in the digestive system.

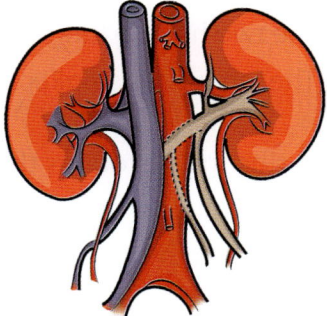

Kidneys

The human body has two kidneys. The kidneys and bladder are part of the excretory system. They filter the blood and remove the waste materials we make as we live and grow. The materials are called urea and ammonia. They are diluted in water and then we excrete them as urine.

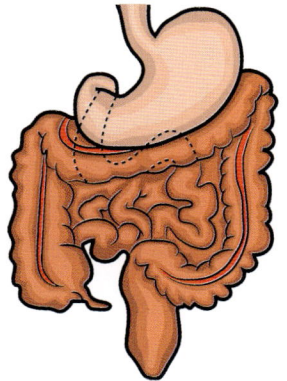

 **Stretch zone**

Research to find out the role of the liver in the body. Write some bullet points to summarise your findings.

Key idea

Each organ and organ system performs a specific task to keep us alive.

■ For more activities, go to Workbook 6 page 47.

Lungs and breathing

In this lesson you will learn about the main functions of the human lungs.

Key words

breathing
exhalation/inhalation
heart
lungs

Think back

What is the function of the lungs? Which gas is vital for life?

We need to take oxygen into our body. This gas is needed to help us break down foods for energy. The foods are broken down into water and carbon dioxide. The carbon dioxide is harmful so has to be sent out of the body.

The process of taking in air containing oxygen and pushing out air containing carbon dioxide is called breathing. This happens in the lungs.

What is energy used for in the body? What would happen if a person ran out of energy?

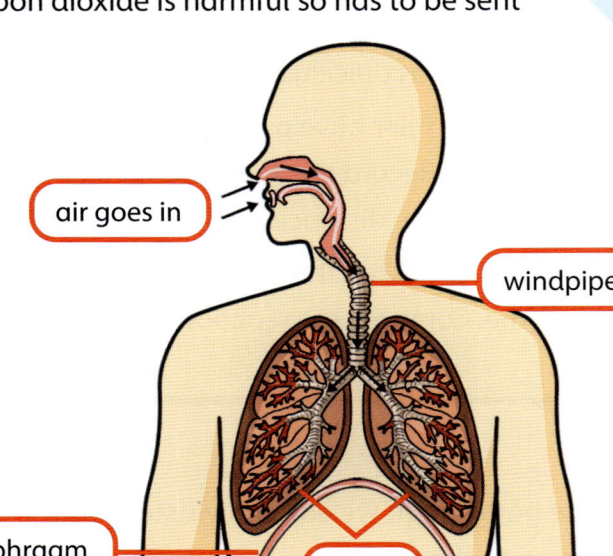

air goes in

windpipe

diaphragm

lungs

Science fact

The two human lungs are different sizes.
The left lung is smaller so there is space for the heart in the chest.

The ribcage surrounds the lungs and heart. When you breathe in, a sheet of muscle called the diaphragm moves down. The muscles between your ribs contract and the ribcage moves upwards and outwards. This makes the space inside your chest bigger. This lets air enter your lungs through your nose and mouth.

Breathing in is called inhalation.

When you breathe out, the muscles between your ribs relax. The diaphragm moves up. This makes the space in your chest smaller and the air is pushed out of your lungs through your mouth and nose.

Breathing out is called exhalation.

What function does the ribcage have in the body?

48

■ For more activities, go to Workbook 6 page 48.

Breathing rates

You are going to investigate rates of breathing.

1 Plan how you can compare breathing rates immediately before and immediately after exercise.

2 List the equipment you will need to use.

3 Carry out your investigation. Think about how you will measure the breathing rates accurately.

4 Record the results for the people in your group.

5 Compare the breathing rates at rest and immediately after exercise. Identify any patterns in the data.

6 Write a short report to describe your investigation. Include your conclusions about why the rates were different.

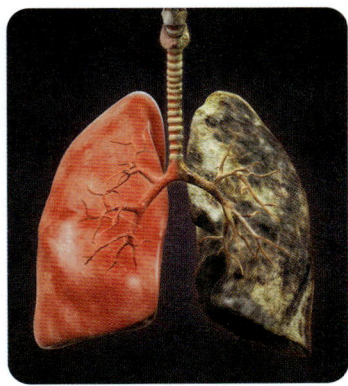

Every part of your body needs oxygen. The oxygen is taken around your body from the lungs by the blood vessels in the circulatory system. This system will be covered in the next lesson.

The lungs are very delicate and can easily be damaged.

This damage can be caused by infection, pollution or smoking.

Talk about examples of when people should wear masks at home, outside the home or at work.

Key idea

The lungs are vital for breathing, the process where oxygen is taken into the body and carbon dioxide is removed.

Stretch zone

Research why the lungs are filled with tiny air sacs and are not one big sac like a balloon. Draw a poster to show your findings.

■ For more activities, go to Workbook 6 page 49.

The human circulatory system

In this lesson you will describe the functions of the human heart, blood vessels and blood.

Think back

Why does oxygen have to be transported to every part of the body?

Key words

artery
blood
blood vessel
capillary
cell
circulatory system
deoxygenated/
oxygenated
heart
lifestyle
plasma
vein

The circulatory system is made up of three main parts: blood, blood vessels and the heart. Its function is to transport blood around the body.

Blood

Blood is a mixture of clear liquid called plasma, red blood cells, white blood cells and cells called platelets.

heart

blood vessels

Human circulatory system

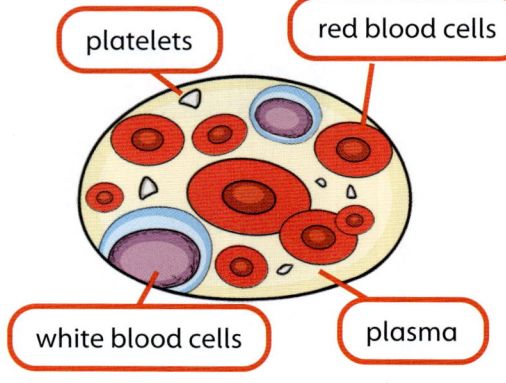

platelets

red blood cells

white blood cells

plasma

Microscope view of blood

Where would you draw the lungs onto this diagram? Why is it important that the lungs have a good blood supply?

The functions of the different parts of blood are shown in the table.

Part of blood	Function
plasma	mainly water, lets the blood flow; it also carries salts, carbon dioxide and nutrients around the body
platelet	helps to stop bleeding
red blood cell	carries oxygen
white blood cell	helps to fight diseases

How is blood suited to its function? Discuss your thoughts with a partner.

■ For more activities, go to Workbook 6 page 50.

Blood vessels

Arteries take blood away from the heart and carry oxygen to the body. They carry oxygenated blood. They have thick walls.

Veins take blood containing carbon dioxide and waste from the body back towards the heart. They carry deoxygenated blood. These have thinner walls and valves to stop the blood flowing the wrong way.

Capillaries are tiny blood vessels connecting arteries to veins. They take oxygen and nutrients to individual cells in the body.

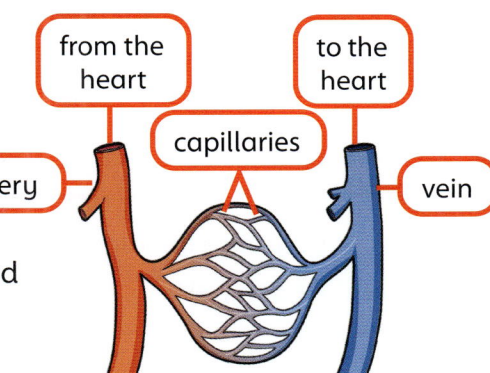

The human heart

The heart is made up of muscle that contracts to pump blood along arteries to the lungs and the body. It has four parts or chambers.

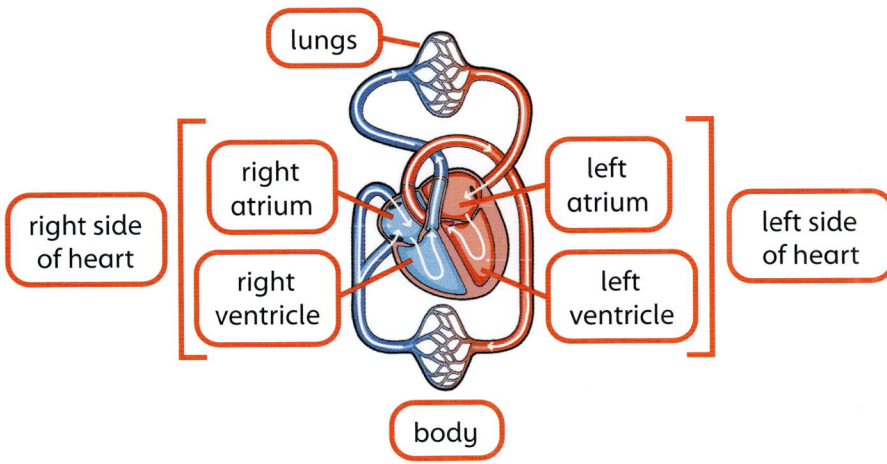

Every time the heart contracts or beats, it pushes a pulse of blood around the body. This pulse can be felt where an artery passes over a bone. A pulse is usually measured inside the wrist.

Problems with the heart

People can have problems with their heart such as high blood pressure, narrowing and blocking of arteries and heart attacks. Doctors can provide medicine or carry out operations to help with these problems. A healthy lifestyle will also help.

Stretch zone

Find out how the valves in veins stop our blood flowing in the wrong direction. Tell your partner how the valves work.

Science fact

If you could stretch out your blood vessels they would reach for over 9500 km! That is longer than the distance between Egypt and Indonesia.

Why does the heart pump blood to the lungs before it is sent around the body?

Key ideas

- The blood is pumped along blood vessels by the heart.
- Blood contains cells and plasma that carry vital materials around the body.

2 Organs and Systems

51

■ For more activities, go to Workbook 6 page 51.

The digestive system

In this lesson you will learn how foods are broken down in the digestive system.

Key words

digestion
digestive system
enzyme
nutrients

Think back

Draw a diagram of the digestive system, label it and write in the function of each part.

Nutrients are needed in the body for life processes. We obtain these nutrients by eating and drinking.

Nutrients are often too large to pass into the body through the blood. They have to be broken down into smaller parts first. This is called digestion and it takes place in the digestive system.

Digestion starts in the mouth with the chewing and mixing that take place during eating. You may remember studying teeth before. Teeth chop up the food – different-shaped teeth do different jobs. Some are sharp like scissor blades to cut into foods. Others are pointed to bite into foods. Some are flat to grind foods.

Which type of tooth is good for biting food? Which is best for grinding food? Discuss how to look after your teeth and why this is so important.

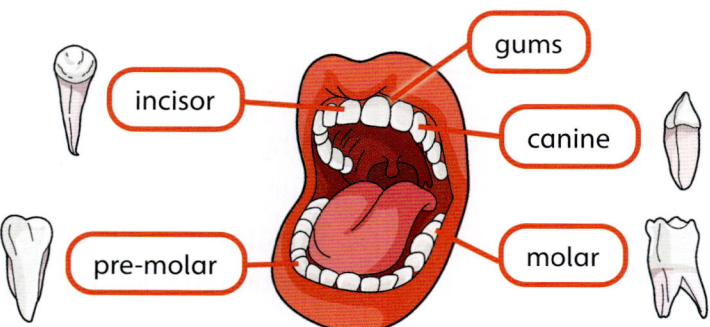

The breakdown of food in the digestive system is helped by chemicals called enzymes. Each enzyme breaks down a particular food type.

In the mouth, saliva contains an enzyme called amylase. Amylase breaks down starch into sugar. Starch is found in foods such as bread, rice and pasta.

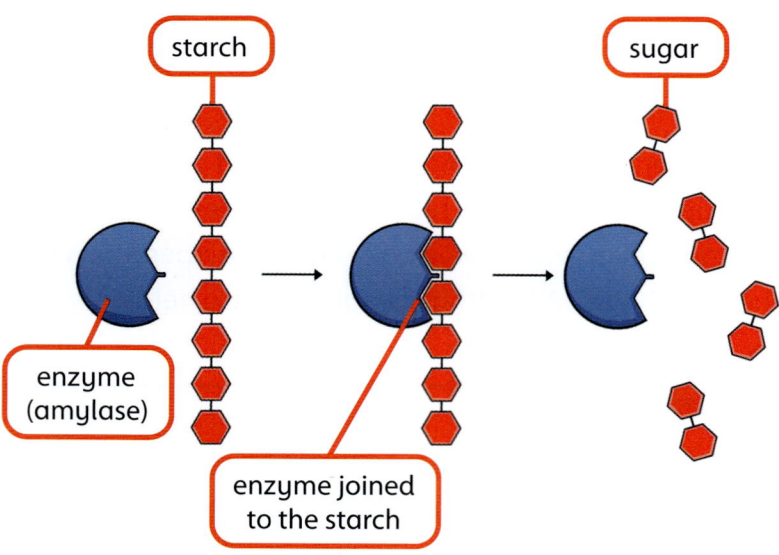

■ For more activities, go to Workbook 6 page 52.

Investigating amylase

You will investigate how amylase in saliva breaks down starch. Starch reacts with a chemical called iodine to give a purple colour. This can be used to test for starch.

1. Plan how you are going to find out if chewing bread helps to break it down from starch to sugar.

2. Make a list of the equipment you will need.

3. Think about how you will make it a fair test. Which variables will you keep the same?

4. Carry out your investigation and record your results in a table.

5. How did the bread change in taste during the chewing?

6. Produce a report of your investigation. This can be a poster, a leaflet or a computer slide show.

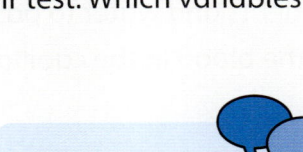
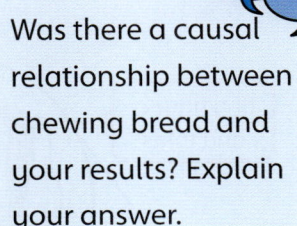

Was there a causal relationship between chewing bread and your results? Explain your answer.

Warning!
Wear safety glasses. Always wash your hands before and after touching the bread. Only handle your own bread.

Be a scientist

Scientists look for causal relationships. They ask: did one thing (the cause) make something else happen?

▶ page 9

Chewed food is swallowed and passes through the oesphagus to the stomach. Here it is mixed with acids and enzymes to form a soft paste. The paste passes from the stomach to the small intestine.

In the small intestine, bile is added. This comes from the liver and helps to break down fats and oils. Enzymes are also added from the pancreas. These break down other fats, and proteins and sugars, as the food is squeezed along the small intestine. The nutrients from the food are now ready to enter the blood.

Finally, the parts of the food that have not passed into the blood are passed to the large intestine. In the large intestine, water enters the body. Waste food then passes out of the body into the toilet.

Stretch zone

Research a reptile or bird to find out how their digestive system compares to a human. Write a report of your findings.

Key idea

Nutrients are broken down in the digestive system into smaller parts that can enter the blood.

2 Organs and Systems

53

Absorbing nutrients and water

In this lesson you will describe how nutrients and water are absorbed into the body from the digestive system.

Key words

absorption
nutrients
re-absorption
surface area
water

Think back

Can you remember where nutrients are absorbed into the body?

In the small intestine, the digested nutrients are small enough to enter the blood. The small intestine is very narrow so food squeezes through it slowly. This gives time for the nutrients and water to pass through the wall of the small intestine into the blood in the capillaries. This process is called absorption.

Science fact

The walls of the small intestine are lined with tiny fingers and folds. This increases the surface area. This means that a lot more nutrients and water can pass through its walls.

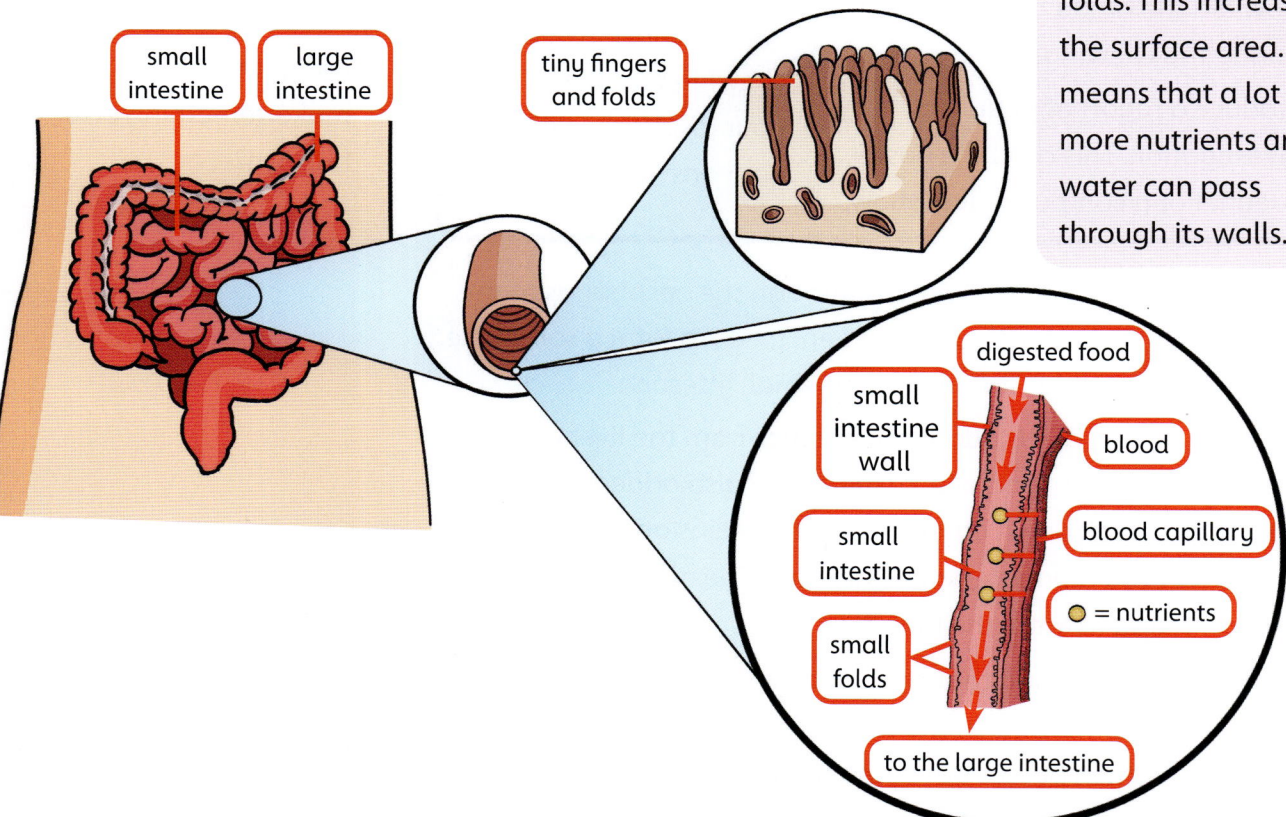

The small intestine has a very good blood supply. The nutrients and water are absorbed into the many capillaries near the wall and are then transported around the body in the blood.

Discuss how the small intestine is adapted to help it to do its job.

■ For more activities, go to Workbook 6 page 54.

Investigating surface area and absorption

You are going to investigate how surface area affects the rate of absorption.

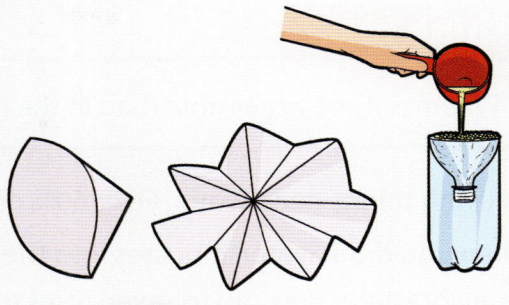

1. Take two pieces of filter paper or paper towel. Fold one into quarters to make a cone. Fold the other into a cone with pleats. Use the diagram to help you.

2. Place the cones into two filter funnels. You can use a filter funnel and a beaker or a cut-up plastic bottle as shown in the diagram.

3. Pour a mixture of soil and water into each cone.

4. Time how long the water takes to pass through each cone and record your results in a suitable way.

5. What do your results tell you about surface area and the rate of absorption? Write a conclusion for your investigation.

Be a scientist

Scientists control their investigations to make them fair. Pour the same volume of mixture into each filter funnel.

▶ page 9

How does this investigation help you to understand why the wall of the small intestine has a lot of fingers and folds?

Once nutrients have been absorbed, the waste food passes from the small intestine into the large intestine. It then passes out of the body into the toilet.

In the large intestine, some more water is taken back into the body. This process is called re-absorption.

If too much water is re-absorbed, the waste passing out is too solid. This is called constipation.

If too little water is re-absorbed, the waste passing out is very runny. This is called diarrhoea.

Stretch zone

Research how the following animals manage a balance of water in their bodies by taking in (absorbing) and removing (excreting) water: frogs, fish, birds and elephants.

Key ideas

- Nutrients and water are absorbed into the blood through the walls of the small intestine.
- Water is re-absorbed into the blood through the walls of the large intestine.

2 Organs and Systems

■ For more activities, go to Workbook 6 page 55.

Water transport and the urinary system

In this lesson you will learn how water is transported and excreted.

Key words
bladder
excretion
kidneys
urinary system

Think back

Where is most water absorbed in the body?

All living things need water. Over 60% of your body is water.

When you drink water, it passes into the stomach and then into the small intestine. Here, up to seven litres a day are absorbed into the blood. A further two litres can be re-absorbed in the large intestine.

Waste materials not needed by the body are removed. This process is called excretion.

You have seen how waste food is sent out of the body after digestion. You have also studied how the lungs remove carbon dioxide. Kidneys also have an important part to play in removing waste.

The kidneys are part of the urinary system. They filter the blood. They are made up of millions of tiny filtering cups. The filters only let small waste materials through. This means they can remove harmful chemicals, such as ammonia and urea, but proteins and blood cells stay in the blood.

How much water do you drink every day? How do you feel if you do not drink enough water?

makes saliva

helps the body to keep cool

plasma in blood carries oxygen and nutrients

helps all reactions in the body such as digestion

protects organs

keeps joints supple

The functions of water in the body

Why is it important that proteins and blood cells cannot pass through the filter?

■ For more activities, go to Workbook 6 page 56.

The waste materials are diluted in the kidneys with water to make urine. The urine is passed down tubes called the ureters to be stored in the bladder. Urine is then sent out of the body through the urethra when a person or animal goes to the toilet.

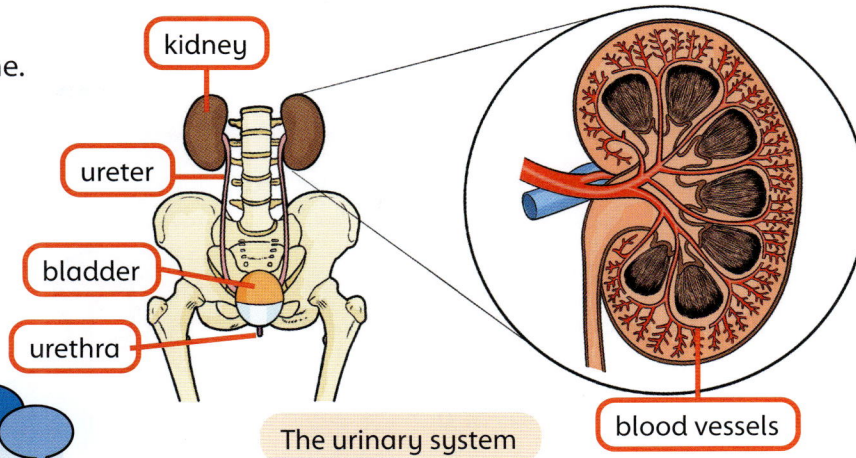

The urinary system

kidney
ureter
bladder
urethra
blood vessels

Why do the kidneys need a very good blood supply?

 Making model kidneys

You are going to design and build a model to show how the kidneys work.

1 Use the diagram to give you some clues.

2 Filter the waste sample your teacher gives you.

3 Present your model to the rest of the class. Explain how it works and how it models the working of real kidneys.

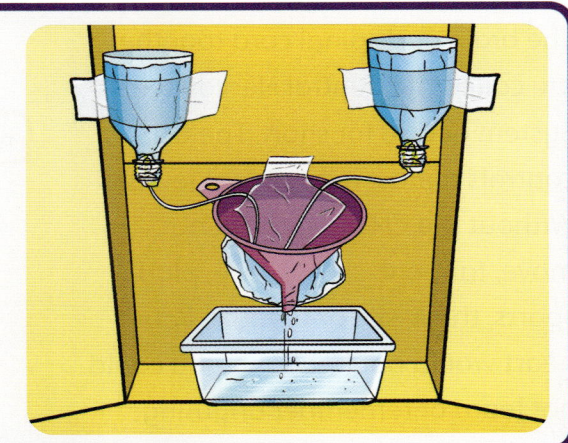

The kidneys can be damaged by injuries such as crashes and falls. They can also stop working well because of an illness. People with high blood pressure can also have problems with their kidneys. When kidneys stop working it is known as kidney failure.

Kidney failure means harmful materials cannot be removed from the blood. The person has to have dialysis. This means their blood is passed through a machine that filters the blood for them. Eventually, a person may need to surgically receive a new kidney from another person. This is called a transplant.

 Stretch zone

Research how excretion happens through the skin. Write a short report of your findings.

Key idea

Water is essential for living things. It is absorbed into the body and is excreted along with waste materials in the urinary system.

2 Organs and Systems

■ For more activities, go to Workbook 6 page 57.

The brain and the nervous system

In this lesson you will learn how the nervous system functions.

Key words
brain
nerves
nervous system
spinal cord
touch

Think back

How are the brain and spinal cord protected?

The brain contains billions of nerves. These link with other nerves in the body either directly or through the spinal cord. The brain sends and receives messages along the nerves. This allows it to control the workings of the body.

The brain and spinal cord together are called the Central Nervous System or CNS for short. The CNS controls almost everything you do. Without the CNS you could not move, talk, eat, sing, dance, play sports, think, see or breathe. Your heart would not beat and you could not learn or remember anything. Nerve endings in your skin would not be able to sense pressure, pain or heat.

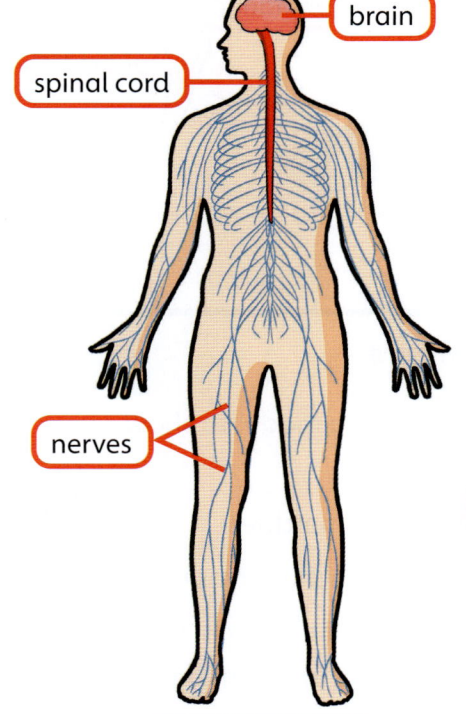

brain

spinal cord

nerves

The human nervous system

Science fact

Messages pass along nerves as electrical signals. They travel at over 250 kilometres per hour! This is why if you touch something hot, you know about it instantly.

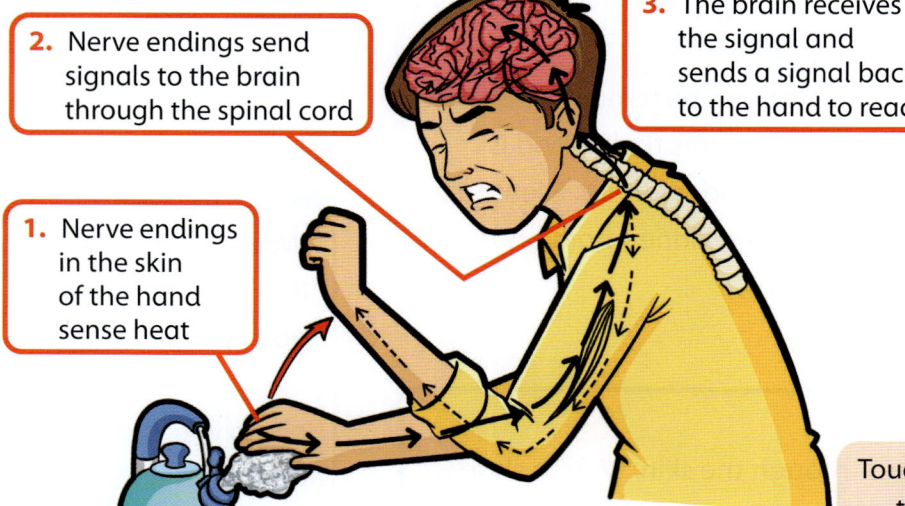

2. Nerve endings send signals to the brain through the spinal cord

3. The brain receives the signal and sends a signal back to the hand to react

1. Nerve endings in the skin of the hand sense heat

Touch is detected by nerve endings that send signals to the brain

■ For more activities, go to Workbook 6 page 58.

Testing your nerve endings

With a partner, you are going to investigate how sensitive the skin is to touch.

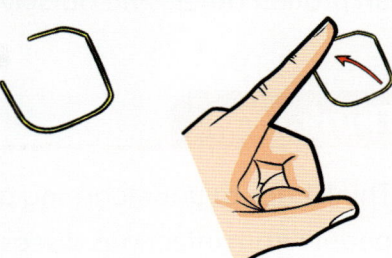

1   Use a bent paperclip with the open ends 2 cm apart (as shown in the first part of the diagram).

2   Ask your partner to look away.

Gently touch the two ends of the paperclip to the finger of your partner (as shown in the second part of the diagram).

Warning! Press very gently. Do not hurt your partner.

3   Ask your partner to tell you if they feel one or two points.

If they say one point: open the paperclips so the ends are a bit further apart and test again.

If they say two points: close the ends slightly and test again.

4   Record the distance when your partner can feel two points.

5   Now investigate the back of the hand and forearm in the same way.

6   Write a report of your investigation. What does your investigation tell you about how far apart nerve endings are in different parts of your body?

Why are the nerves in your skin so important?

Stroke

When the blood supply to the brain stops, the brain cells will die. If this happens a person may not be able to walk or talk properly. This is called a stroke. With medicines and therapy, the person who has suffered a stroke may get better, but often they may not fully recover.

High blood pressure, fat building up in blood vessels, an illness called diabetes and smoking can cause strokes.

Falling over or objects landing on the head can also injure the brain.

How is this person protecting their brain? Discuss some other examples of how people protect themselves from head injuries.

Key idea

The brain and nervous system are essential for controlling many of the body's functions.

■ For more activities, go to Workbook 6 page 59.

Infectious diseases and their prevention

In this lesson you will learn about infectious diseases and how we can protect others and ourselves from them.

Key words

barrier
defence mechanism
host
infectious disease
microorganism
prevention
secretion
transmission
vaccine
vector

Think back

Think back to your learning about vaccines. How do they protect people from infectious diseases?

Microorganisms are all around us. They are in the air, water and soil. Some microorganisms can cause diseases. Disease-causing microorganisms are called pathogens. Some examples are:

- Viruses – such as influenza, COVID-19, measles and smallpox
- Bacteria – such as salmonella food poisoning, anthrax and cholera
- Parasites – such as tapeworm and malaria parasite
- Fungi – such as ringworm and athlete's foot

A person or animal infected with a pathogen is called the host. Pathogens can pass from host to host. This is called transmission and these diseases are called infectious diseases.

Pathogens can enter your body in different ways. Some examples are shown below. Animals that spread diseases are called vectors.

 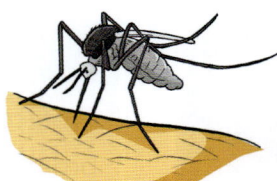  

How diseases could be spread

Where would you prefer to get your drinking water from? Write down some of the problems drinking polluted water can cause.

Look at the pictures. Talk about how diseases could be spread in each one. What do you recommend these people do to prevent diseases spreading this way?

Preventing diseases

Good hygiene can help to stop the spread of diseases.

Clean water is important. Dirty water can be filtered or boiled or purifying chemicals can be added. Food should be washed and fruits and vegetables can be peeled. Cooking foods can kill microorganisms.

Discuss with a partner some of your earlier work on hygiene. List some of the ways that good hygiene can help diseases from spreading.

■ For more activities, go to Workbook 6 page 60.

Researching diseases and pathogens

You are going to work in a small team to produce a poster about some common diseases.

1   Use secondary sources to research a disease caused by:
    a) a virus, b) a bacterium, c) a parasite and d) a fungi.

2   For each disease, find out:
    - The name and description of the pathogen in each example.
    - How dangerous the disease is.
    - How the disease is transmitted (spread).
    - How the disease can be treated or prevented.
    - Where in the world the disease occurs most, and why.

3   Display your poster to make a disease exhibition.

Defence mechanisms

Your body defends itself against infectious diseases. It has physical barriers to stop pathogens entering your body. It also has chemicals, which are in your body's secretions. These are chemical barriers.

Defence mechanism	How it works
Skin	This is a tough covering and is waterproof. It stops pathogens from entering your body.
Sweat	This is a secretion made by the skin and contains chemicals that can kill pathogens.
Tears and saliva	These secrete enzymes that attack pathogens and keep your mouth, throat and eyes safe.
Mucus	This is a sticky secretion that lines your nose, throat and into your lungs. This traps pathogens and stops them entering too far into your body.
Cilia	These are tiny hairs that line the windpipe. These move and set up waves away from your lungs. This moves any pathogens up towards your nose and mouth so you can get rid of them.
Stomach acid	This secretion destroys pathogens you have swallowed.

Study the table. Identify the physical barriers and the chemical barriers. Make a poster that explains to people how their body has barriers to protect them from infectious diseases.

Key ideas

- Microorganisms can cause infectious diseases. These spread from host to host in many ways.
- We can prevent diseases by acting sensibly. Our body also has its own defence mechanisms.

■ For more activities, go to Workbook 6 page 61.

A healthy diet

In this lesson you will recognise the impact of your diet on the health of your body.

Key words

diabetes

diet

health

heart disease

obesity

Think back

Why do people need to eat food? What types of food are there?

How healthy is the meal in the photograph? What might happen to a person who only ate this type of food? What is missing from this meal?

Humans need to eat a balance of different nutrients to stay healthy. This is called a balanced diet. The types of nutrients are shown below.

Nutrient	Importance to the body
proteins (meat, fish and plant alternatives such as lentils and beans)	needed to make muscles and enzymes
carbohydrates (bread, cereals, pasta and potatoes)	used for energy
fats and oils (dairy products, fried foods and creams)	used for energy and insulation
minerals and vitamins (fresh fruit and vegetables)	needed for healthy growth, for example calcium in bones

Study the healthy eating plate below.

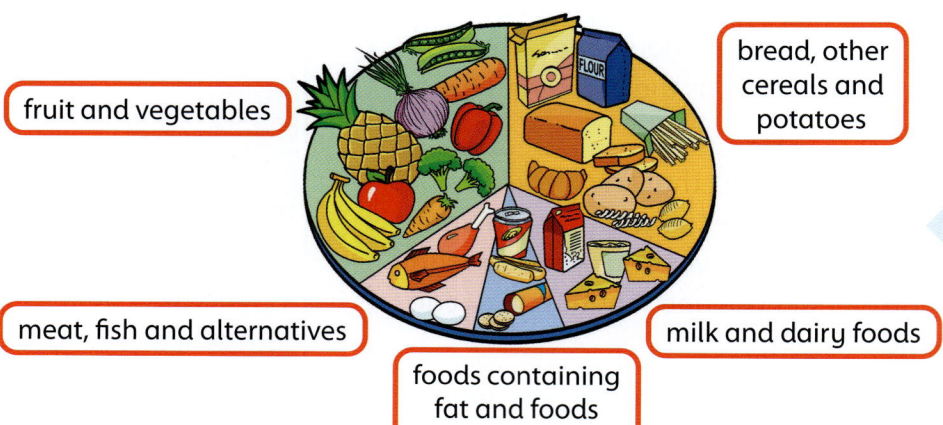

fruit and vegetables

bread, other cereals and potatoes

meat, fish and alternatives

milk and dairy foods

foods containing fat and foods containing sugar

Use the plate to help you to list three sources for each of the following nutrients: protein, fats, carbohydrates and vitamins.

62

■ For more activities, go to Workbook 6 page 62.

How diet can affect organs and systems

1 Use secondary sources to research how a person's diet can cause damage to each organ/organ system you learned about in previous lessons.

2 Design a table to record your research.

3 Display your findings in class.

We need energy but if we eat too much fat and sugar and do not use up the energy then it is stored as fat in the body. Too much fat is unhealthy and causes problems such as obesity. This means a person is very overweight. Obesity causes other health problems such as diabetes, heart disease and joint problems.

Food labels survey

To find out which are healthy and less healthy foods you are going to carry out a food labels survey.

1 Study the labels of the food cans, bottles and packets you are given.

2 Classify the foods into four types:

- high in protein
- high in carbohydrate
- high in fat
- high in minerals and vitamins.

3 Design and complete a table to record your results.

4 Use your table to plan a healthy meal. Produce a menu card to share with your family and, if possible, cook them the meal.

Be a scientist

When scientists use information from other sources rather than doing an investigation themselves, it is called using secondary sources.

▶ page 8

Science fact

A small can of fizzy drink can contain up to 12 teaspoons of sugar. That is 180 calories but no valuable nutrients.

Key ideas

- Animals and humans need to eat a balanced diet of the correct amount of different nutrients.
- Eating too much fat and sugar can cause obesity and other health problems.

■ For more activities, go to Workbook 6 page 63.

Healthy life choices

In this lesson you will recognise the impact of lifestyle choices on health.

Key words

drug

exercise

hygiene

medicine

smoking

Think back

What life choices can people make that can either help to keep them healthy or possibly make them unhealthy? Write a list.

Choosing to eat a balanced diet and to exercise are healthy life choices.

Exercise

What are the people in the photograph doing? Why do they need to eat a healthy diet to be good at their sport?

Science fact

Scientists recommend that you exercise for at least an hour per day across the week. This will help your movement skills and will strengthen your body.

Exercise survey

1 Plan a survey to find out which types of exercise the people in your class do.

2 Ask each person to tell you the exercise they do most often. Find out how long they do the exercise for.

3 Design and complete a table to record your findings.

4 Present your findings as a chart. You will need to decide which is the best chart to use for the data.

5 Use the data to conclude what the most popular type of exercise is and what the average time spent on exercise per week is.

6 Design a poster or presentation to share your findings. Include some drawings or photographs to show examples of the exercises.

Do you think your classmates do enough exercise? Use data to validate your thoughts.

■ For more activities, go to Workbook 6 page 64.

Drugs

Some chemicals have an effect on the body when they are taken. They might take away pain or make a person feel happy or more confident. These chemicals are called drugs.

When a doctor uses drugs like these, they are called medicines. Medicines have been tested to check they are safe and must be taken in the correct amounts. These are called prescription drugs. Examples are antibiotics for infections and steroids for asthma.

People can buy some medicines to treat themselves. These have to be used carefully by following the instructions. These are called 'over the counter' drugs. Paracetamol and ibuprofen are examples.

Some people take drugs in large amounts without the advice of doctors. These can damage a person's health. Using a drug in this way may be called drug abuse. Many of these drugs are illegal.

Stretch zone

Research the dangers of smoking. Produce a poster to persuade people not to smoke. An example is shown opposite. Include examples of diseases caused by smoking.

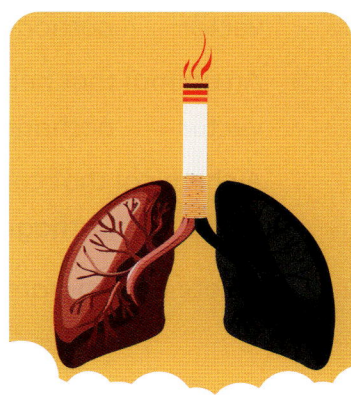

WORLD NO TOBACCO DAY

Good hygiene

It is important to keep our body and clothes clean. This is called hygiene. This helps to prevent diseases and prevents unpleasant smells.

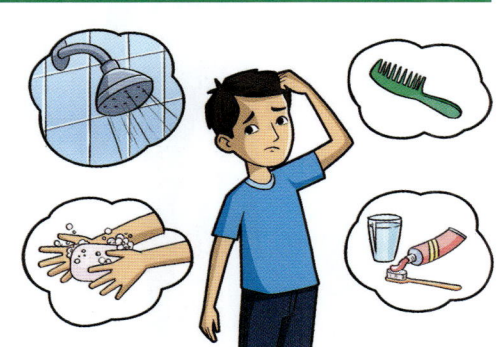

Discuss the examples of good hygiene shown in the picture. Why is each one important? Think of one other thing you should do to make sure your personal hygiene is good.

Key ideas

- Some people make unhealthy life choices, and this can affect their bodies.
- Eating a balanced diet, exercising and having good personal hygiene are healthy life choices.

Check how much you know.
Try the questions on pages 66–67.

■ For more activities, go to Workbook 6 page 65.

1 Circle the healthy life choices from the list below. Put a line through any unhealthy lifestyle choices.

> over-eating   taking exercise   sitting watching TV all day   washing hands   smoking
> eating a balanced diet   brushing teeth   eating sugary foods   eating fruit and vegetables
> abusing drugs   not washing clothes   wearing the same clothes every day
> taking a shower or bath   eating a lot of fried foods   not eating fruit and vegetables

2 Circle the correct answers.

a A microorganism that causes a disease is called a:

vector   pathogen   vaccine   medicine

b An animal that helps to spread a disease is called a:

vector   pathogen   vaccine   drug

c A chemical that contains dead or damaged microorganisms to help prevent disease is called a:

vector   pathogen   vaccine   drug

3 Complete the diagram with the names of the main organs.

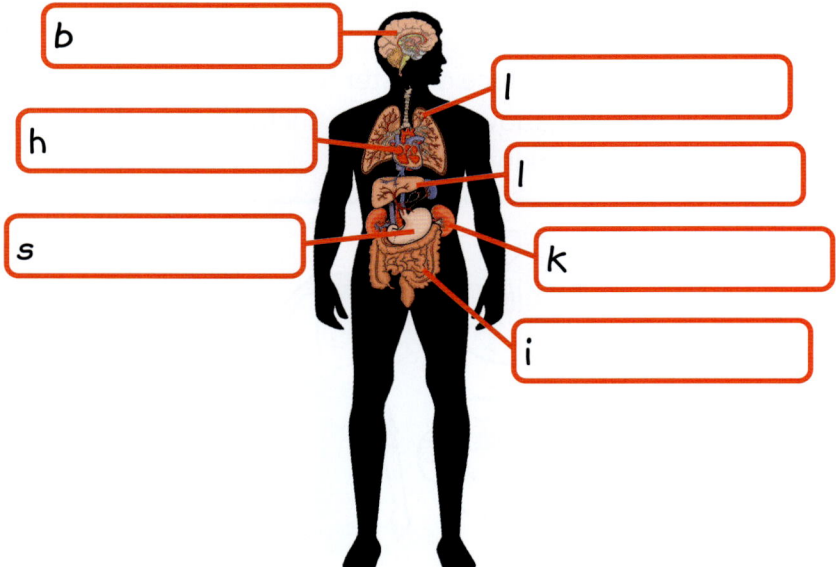

b _____

h _____

s _____

l _____

l _____

k _____

i _____

4 a Label the diagram of the urinary system.

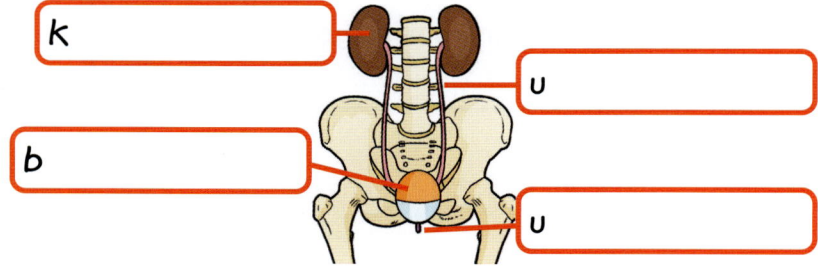

k _____

b _____

u _____

u _____

b What is the function of the urinary system?

_____

■ For more activities, go to Workbook 6 page 66.

5 Label the diagram of blood vessels.

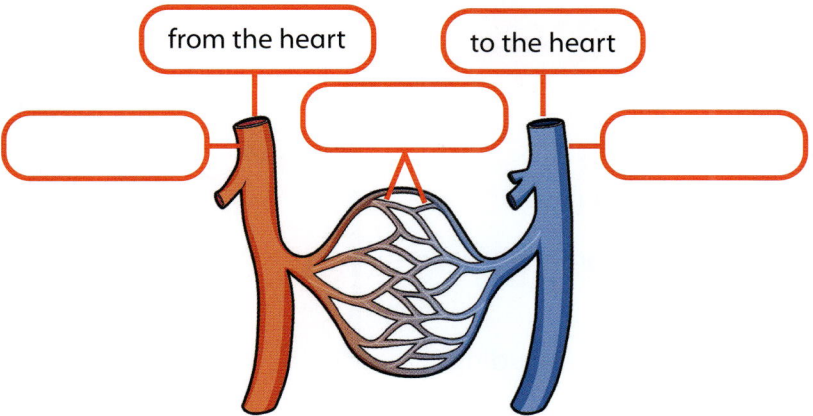

6 Complete the table.

Nutrient	Importance to the body
	needed to make muscles and enzymes
carbohydrates	
fats and oils	
	needed for healthy growth, for example calcium in bones

7 Study the table. It shows the results of two people who had their resting pulse rate taken and then their pulse rates taken at different times after exercise. All the rates are in beats per minute.

Person	Resting pulse rate	Pulse rate just after exercise	Pulse rate 30 seconds after exercise	Pulse rate 60 seconds after exercise	Pulse rate 90 seconds after exercise
A	75	140	125	98	84
B	80	142	130	110	90

a Which person had the lowest resting pulse rate? _____

b Which person showed the largest increase in pulse rate after exercise? _____

c Which person had the fastest recovery rate? _____

d Why did both people show an increase in pulse rate after exercise?

_____

_____

e Describe one variable you would keep the same in this investigation to make it a fair test.

_____

2 Organs and Systems

■ For more activities, go to Workbook 6 page 67.

3 The Way We See Things

In this unit you will:

- recognise that light appears to travel in straight lines
- revise that we see light sources because light from the source enters our eyes
- explore how light can be reflected from surfaces, including mirrors
- describe other properties of light, such as refraction and how we see colour
- investigate shadows to understand their shape and size
- discover how light is measured.

> beam   light intensity   light source   mirror   opaque   ray   reflect   shadow   silhouette   translucent   transparent

The Sun is our main source of natural light. The Sun gives off so much light it can heat and light up whole planets.

Look at the photograph. Find any places where light is bouncing off or reflecting from surfaces. How do you use reflections?

Science fact

Before televisions and cameras, people used to pay to have portraits painted. In the 1700s, Étienne de Silhouette invented a new way of making pictures of people. He invented the silhouette. These were much cheaper to produce than a painting.

How is the artist using a shadow?

Why does the room have to be darkened?

Why is it better to use a lamp than a candle?

The people in this street know that there is not a giant nearby! Why are the shadows so large?

The people in the photograph are enjoying the sunshine.

Why will some people have to move in order to stay in the Sun for longer?

■ For more activities, go to Workbook 6 pages 68–69.

Sight

In this lesson you will explore our sense of sight and how we use it to learn about the world around us.

Key words

eyes

sense

sight

Think back

We need light to see. Light enters our eyes to let us see the world around us.

Like all our senses, our sight tells us about our surroundings.

Some animals have eyes at the side of the head, like this horse. This helps them see if there is any danger around them.

Humans, and many animals, have eyes at the front of their head. This helps us see things in front of us in a lot of detail.

Science fact

About 95% of animals have eyes. Spiders usually have eight eyes.

Is seeing with two eyes better than one? Part 1

You are going to investigate if you need two eyes to see.

1. Hold a pencil in each hand and stretch your arms out.
2. Close only your left eye and try to make the ends of the two pencils meet.
3. Ask a partner to measure and record how far away the pencils are from your eyes.
4. Now close only your right eye and try to make the ends of the pencils meet.
5. Again, ask a partner to measure and record how far away the pencils are from your eyes.
6. Now try this again with both your eyes open.
7. Ask a partner to measure and record how far away the pencils are from your eyes.
8. Was there a link between the distance and which eye or eyes you had open? Write a report of your investigation.

What happens? Is it easy to make the pencils meet?

Does this make a difference? Is it easier to make the pencils meet with both eyes open?

70

■ For more activities, go to Workbook 6 page 70.

Is seeing with two eyes better than one? Part 2

1. Sit your partner at a desk. Put a paper cup on the desk about an arm's length away from them.

2. Ask your partner to close their left eye.

3. Hold a paperclip in your hand. Move the paperclip slowly over the cup about an arm's length above it.

4. Your partner shouts 'drop it' when they think the paperclip is above the cup and will fall directly into it.

5. Carry out the test until you have dropped ten paperclips. Record how many land in the cup.

6. Repeat with your partner, first closing their right eye, and then having both eyes open.

7. Carry out the test until you have dropped ten paperclips. Record how many land in the cup.

8. Swap roles and repeat the investigation.

9. Analyse your results and present your results as a table and a chart.

Does using two eyes make it easier to get the paperclip into the cup?

Did it make it easier for everyone? Compare your findings with others.

Reading with one eye closed

1. Close your left eye and point to a word.

2. Open your left eye and close your right eye.

3. Now look at the word with both eyes.

What happens to the position of your finger?

What have you discovered in your investigations? Is it better to have two eyes or one? Discuss your ideas as a class.

Stretch zone

Research animals that have eyes in different places on their heads, such as hawks, flat fish and chimpanzees. Find out what difference this makes to how they see things and how this helps them to live in their habitat.

Key idea

Light enters our eyes to let us see the world around us.

■ For more activities, go to Workbook 6 page 71.

Brain tricks

In this lesson you will explore optical illusions.

Key words

blind spot
optical illusion
sight

Think back

When there is no light, it is dark. How well can you see in the dark?

Science fact

Hawks and eagles have very good sight. It can be eight times better than human sight. A hawk can see a rabbit 1.5 kilometres away.

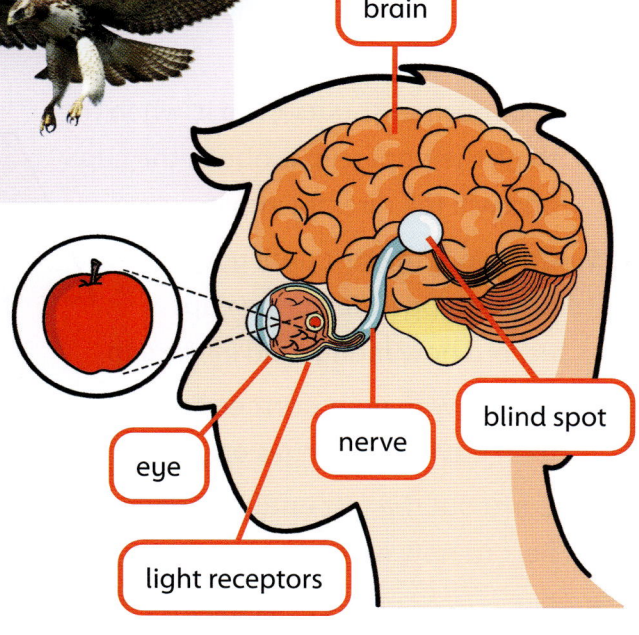

brain

blind spot

nerve

eye

light receptors

Our sight is not as good as lots of animals.

To see we need light. The light from the object we are looking at hits the lining at the back of our eye which has special nerve cells called receptors.

These receptors respond to light and send a message to our brain. The brain works out what we are looking at. One part of this lining does not have any receptors and is called the blind spot. If the light from an object hits this part of the eye, we cannot see it.

How can we show that we have a blind spot?

The blind spot

1   Make a card like this one.

2   Hold the card in front of you. Close your right eye.

3   Stare at the X with your left eye. Slowly move the card closer to your face.

4   Keep looking at the X with your left eye. The dot will disappear. You have found your blind spot.

5   Repeat this test with the other eye. This time close your left eye and find the blind spot in your right eye.

6   Use the diagram of the eye above, and the results from your investigation, to produce a small poster that explains to people what a blind spot is and why we do not usually notice it.

72

Our eyes work very closely with the brain. Sometimes the brain plays tricks on us. You may have heard of a mirage. This is when we think we have seen something, but it isn't really there.

We can play games to try to trick the brain.

Brain tricks

1 Make a card like the one in the diagram.

2 Hold the card in front of you.

3 Close your left eye.

4 Stare at the cross and move the card closer to your face.

5 What happens to the space in the middle of the lines?

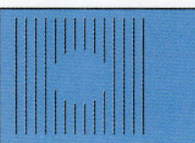

Be a scientist

When observing investigations, scientists have to be careful their brain isn't tricking them into seeing things that are not there. That is why they prefer to use measuring devices.

▶ page 10

> What happened to the space for everyone else? Did you all see the same thing?

In the investigation, our brain recognises that something is missing. It thinks it is the blind spot and fills the space in for us.

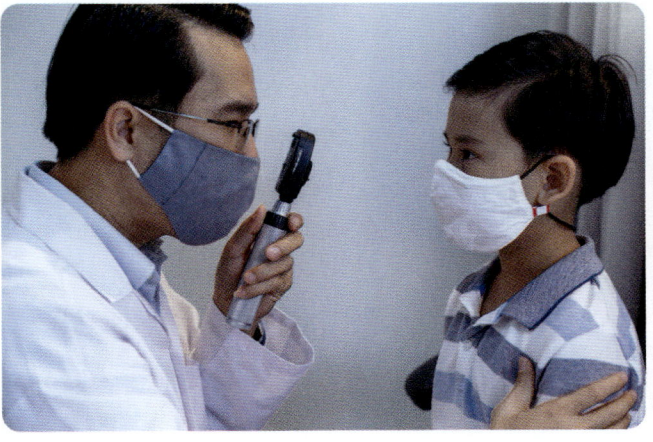

An optician checks that our eyes are working properly. If they are not, they can work out what kind of glasses we need

> Have you ever had your eyes tested? Talk to a partner about what happened.

Key idea

Our eyes send messages to our brain. Sometimes our brain tricks us into seeing things that aren't there.

Stretch zone

Use the internet to find out some of the ways that devices can be used to improve a person's sight. Draw some examples.

■ For more activities, go to Workbook 6 page 73.

How do our eyes see things?

In this lesson you will recap that we can see because light from a light source enters our eyes.

Key words
light source
mirror
reflect

Think back

Think back to how light travels from a source to our eyes. List three light sources that you have seen today.

Some of the light sources are natural. The Sun is a good example. Other sources of light come from objects that people have made. These are called artificial sources of light.

When light is bounced back from an object, we say it has been reflected. An object that does this well is called a good reflector; an object which does this badly is called a poor reflector.

Testing for good reflectors

You are going to use an artificial source of light to test whether some objects are good reflectors. The light source is a torch.

Find objects around the room to investigate.

1   Choose an object to test.

2   Shine the light from the torch onto the object. Write down whether light bounces back off the object or not.

3   Test some other objects in the same way.

4   Look at your observations. Group the objects into good reflectors and poor reflectors and design a results table.

Predict if the object will reflect light.

Choose the object that you think is the best reflector out of the ones you tested.

What kind of material is it? Describe the properties of the material to your partner. Would you describe the material as shiny or dull?

Be a scientist

Remember that scientists do not just guess when they make a prediction. They use their scientific knowledge to work out what they think is most likely to happen.

▶ page 8

If something is shiny, it reflects light because it is very clean or polished. If it is dull, it is not bright or clean and does not reflect light.

Remember: There has to be a source of light for us to see things.

■ For more activities, go to Workbook 6 page 74.

Think back

The Moon is not a source of light. It is like a giant mirror in the sky. The Moon does not make its own light. It reflects it from the Sun.

Can a mirror make its own light?

Your teacher will give you a small mirror.

1 Look in the mirror. What do you see?

2 Hold your hands around the mirror to block the light. What do you see?

3 Does this prove that the mirror does not make its own light?

How did the 1969 Moon landing prove to us that the Moon is not a light source?

What do you think it looks like on the far side of the Moon?

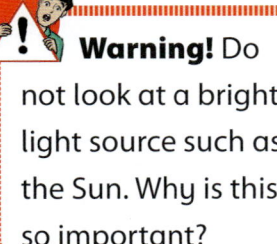

Warning! Do not look at a bright light source such as the Sun. Why is this so important?

If we look at a light source, the light is travelling directly into our eyes.

When we look at objects that are not light sources, we see them in a different way. The light is not made by the object. It is reflected from the object to our eyes.

Key idea

Without light, our eyes cannot see anything.

■ For more activities, go to Workbook 6 page 75.

The journey of light

In this lesson you will explore that beams or rays of light can be reflected by surfaces, including mirrors, into our eyes.

Key words

beam

mirror

ray

reflect

Think back

Remember, light is reflected from some surfaces better than others.

Discuss the different uses of mirrors you can see in the photographs. Talk about examples of mirrors you have used.

Science fact

The image you see of yourself in a mirror is back to front (the wrong way round). Test this!

■ For more activities, go to Workbook 6 page 76.

Can you remember the types of material that reflect light? Discuss two examples.

The mirrored image: Part 1

Your teacher will give you a small hand mirror and a larger mirror.

1   Investigate how to use the mirrors to view the back of your head.

2   Is this a good use of mirrors?

3   How did the mirror image of the back of your head differ from your actual head?

The mirrored image: Part 2

1   Place an object in front of a mirror. Point to the left side of the object and look in the mirror. Record what you see.

2   Point to the right side of the object and look in the mirror. Record what you see.

3   Draw two large copies of the star shown below. Label one A and the other B.

A

B

4   Try to draw a pencil line between the two outline lines of star A. Do not touch the edges.

5   Now collect a small mirror. Look at star B in the mirror and try to draw a pencil line between the two outline lines. Do not look at star B, only at its mirrored image.

Look at the results from your investigations.

Do your results support the Science fact on the opposite page?

Key ideas

- Some surfaces, such as mirrors, reflect light.
- Mirrors have many important uses.

■ For more activities, go to Workbook 6 page 77.

3 The Way We See Things

Uses of reflection

In this lesson you will explore that light can be reflected by surfaces, including mirrors, into our eyes and we see the object.

Key words
mirror
periscope
reflect

Think back

How did you use mirrors to see behind you?
What happened to the image?

Mirrors are usually made out of ordinary glass or coated plastic. The back of the mirror is covered in a shiny metal foil that is designed to reflect light very well.

Mirrors are really good reflectors. Many other shiny objects reflect light well.

Dark and black objects do not reflect light very well. This is because the light beam cannot bounce off the surface as well. Instead it is absorbed by the object.

Mirrors are very useful. They help us to see things that we normally cannot. People who drive vehicles use mirrors to see behind them.

Why are mirrors used on the inside and the outside of buildings?

How can reflective surfaces be a problem?

Discuss some important uses of mirrors. Write a list of those that make life safer for us.

■ For more activities, go to Workbook 6 page 78.

Making a periscope

An important use of mirrors is in a device called a periscope. This uses two mirrors to allow us to see round corners or above objects.

You are going to make and test a periscope.

1 Use the diagram to help you to design your periscope.

2 Make a list of the equipment you will need.

3 Make your periscope and test it by seeing if you can see over or around objects.

4 Produce an information leaflet to tell people how to make a periscope and why they are useful. Include a diagram to show what happens to the light from an object to a person's eye.

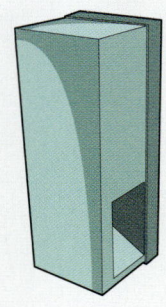 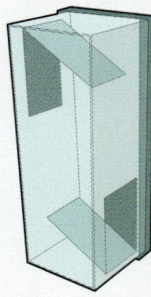

More uses of mirrors

Mirrors have many other uses. Dentists and doctors use them to see into difficult places. For example, a dentist has to be able to see behind your teeth. They are also used to make the light in a torch brighter.

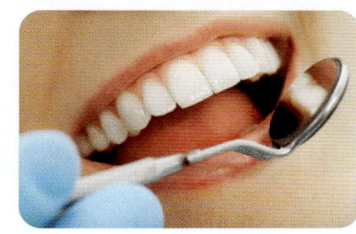

Does foil in a torch make the light brighter?

Your teacher will demonstrate how bright a torch is.

1 Your teacher will shine the torch at a surface. How bright is the light?

2 Your teacher will cover the foil with dark card or paper. Predict what the light will look like now. Do you think it will be brighter or duller?

3 When your teacher switches the torch on now what does the light look like? Did you predict correctly?

4 Write down your conclusions.

Warning! Do not look directly at the torch as the light could damage your eyes.

Be a scientist

Scientists include a scientific explanation in their conclusions. They do not only describe what happened but their ideas about why it happened.

▶ page 12

Stretch zone

Research the use of mirrors that are not flat. They may be curved outwards (convex) or curved inwards (concave). Draw and label some examples.

Key idea

Mirrors help us to see behind us, around objects and over objects.

■ For more activities, go to Workbook 6 page 79.

Ray diagrams

In this lesson you will investigate how a ray of light changes direction when it is reflected from a surface.

Key words

direction

ray

ray diagram

The person in the photograph is seeing the Moon but not looking directly at it. How is light from the Moon travelling to the person's eyes?

Scientists show the direction of light by drawing ray diagrams. The straight lines show the beam or ray of light and the arrows show the direction the light is travelling in.

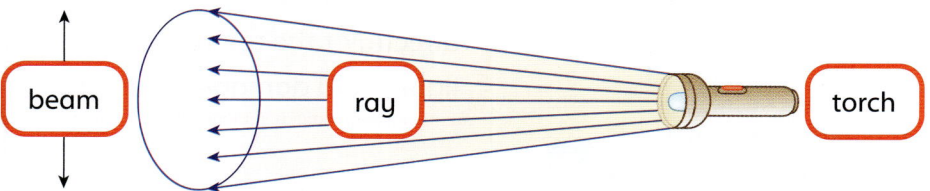

beam

ray

torch

Science fact

A very narrow stream of light is called a light ray. A beam is a lot of rays added together.

If you shine a torch onto a dark surface, you can see the beam of light. The beam leaves the torch and travels through the air until it hits the surface.

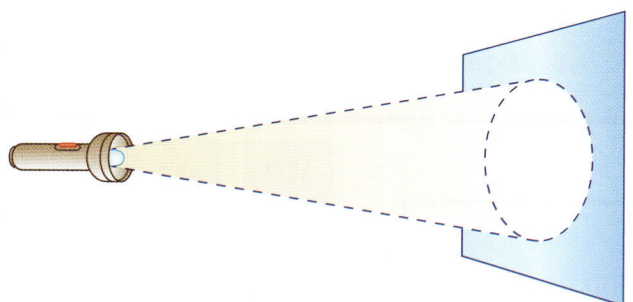

Think back

Remember, light travels in a straight line. You can prove this using the ray box you make next.

Scientists use ray boxes to investigate light. Ray boxes produce a good ray of light to use.

■ For more activities, go to Workbook 6 page 80.

Making a ray box

You can make your own ray box using a torch and a small box such as a shoe box.

1   Use a ruler to find the centre of one of the short sides of your box. Cut a slit from the open side of the box towards the centre.

2   Place your lit torch on the table. Place your ray box bottom side up over the torch so that the slit touches the table surface.

3   Darken the room. Move your ray box until you get a long, thin beam of light shining out through the slit.

4   Shine the light across a piece of paper and draw along the path of the beam. Using a ruler will help you to do this neatly. You have drawn a ray diagram.

5   Now place a mirror in front of the ray of light.

    What happens to the ray of light when it reaches the mirror?

6   Draw a line along the reflected beam of light to show the ray diagram you make when using a mirror.

You should have made a ray diagram like this one.

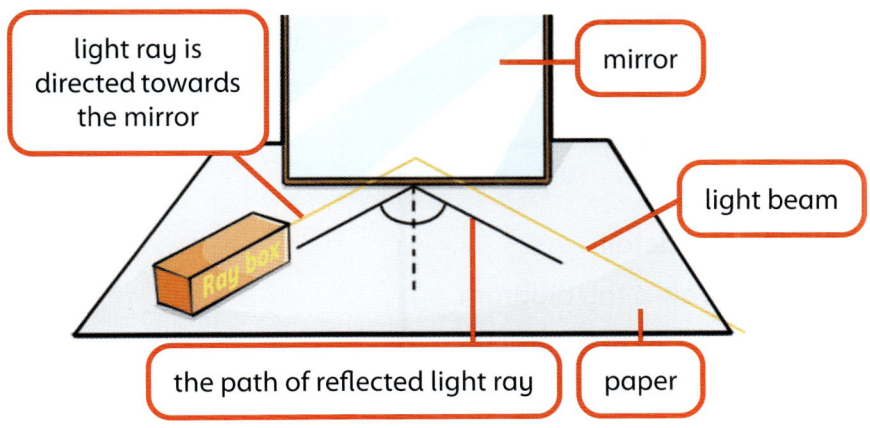

light ray is directed towards the mirror

mirror

light beam

the path of reflected light ray

paper

Using this learning, discuss how light is reflected into your eyes when you look in a mirror.

Key idea

We can follow the journey light takes using ray diagrams.

Stretch zone

Think back to your periscope. Draw a ray diagram to show how a person can use a periscope to see a tree that is hidden by a wall.

■ For more activities, go to Workbook 6 page 81.

Light changing direction

In this lesson you will explore why a beam of light changes direction when it is reflected from a surface.

Key words

angle of incidence
angle of reflection
incident ray
normal
reflected ray

Think back

A beam of light is made up of many light rays. Think back to the ray box you made. When you used the flat mirror, what happened to the ray of light?

Flat mirrors reflect a really good image of objects, called a true likeness. This is because the smooth surface does not scatter the light.

If we stand in front of a mirror, we see ourselves. The light reflected from us will travel to the mirror and be bounced back or reflected into our eyes.

What happens if you look into a mirror from the side? Do you see yourself? What do you see? Where do you have to stand to be able to see yourself?

Mirrors

Use your ray box to investigate mirrors in more detail.

1 Set up the investigation so that you can draw a ray diagram of your ray of light reflecting off the mirror. This time, draw a line to show where the mirror is.

2 Draw three different ray diagrams by moving your ray box. Make sure you know which ray is which. Use a different colour for each one.

3 Take one ray diagram and draw a line at 90 degrees (a right angle) to the mirror. This is called the normal line.

■ For more activities, go to Workbook 6 page 82.

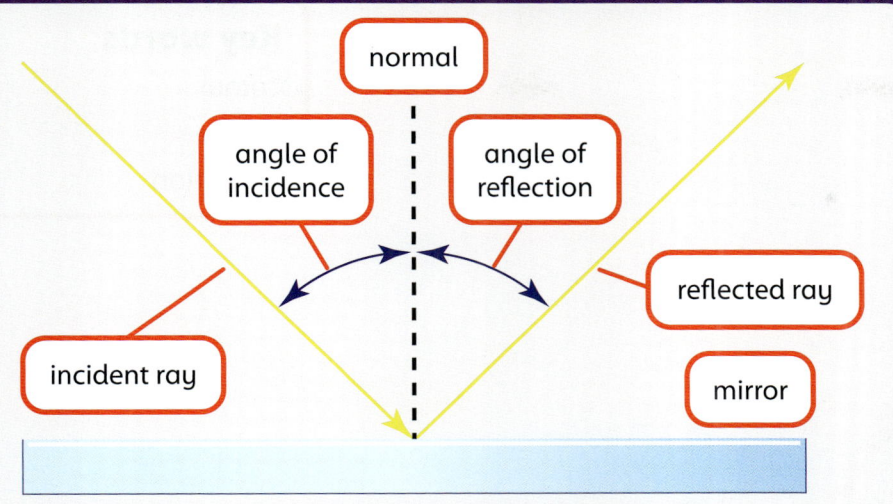

- The light reaching the mirror is the incident ray.
- The angle at which the incident meets the normal line is called the angle of incidence.
- The light leaving the mirror is the reflected ray.
- The angle at which the reflected ray is reflected from the normal line is called the angle of reflection.

4 Label the angle of incidence, angle of reflection, incident ray and reflected ray on one of your ray diagrams.

5 Measure and record the angle between the angle of incidence and the normal line.

6 Measure and record the angle between the angle of reflection and the normal line.

7 Measure and record the angle of incidence and the angle of reflection for your other two ray diagrams.

8 What do your results show?

With a flat, plane mirror the angle of incidence and the angle of reflection are always the same.

angle of incidence = angle of reflection

Stretch zone

Research how you could use the rule above to design a device that lets you see around a corner. Draw your design.

Key idea

Light travels in straight lines but can change direction.

■ For more activities, go to Workbook 6 page 83.

More on light

In this lesson you will describe some other properties of light, such as refraction and colour.

Key words
colour
reflect
refraction

Refraction

The straw is straight, but it looks bent. This is because the light we see is bent. This happens because when light passes from air into a different material, such as water or glass, it changes direction. This is called refraction. This is why things underwater look nearer than they are. This is another type of optical illusion.

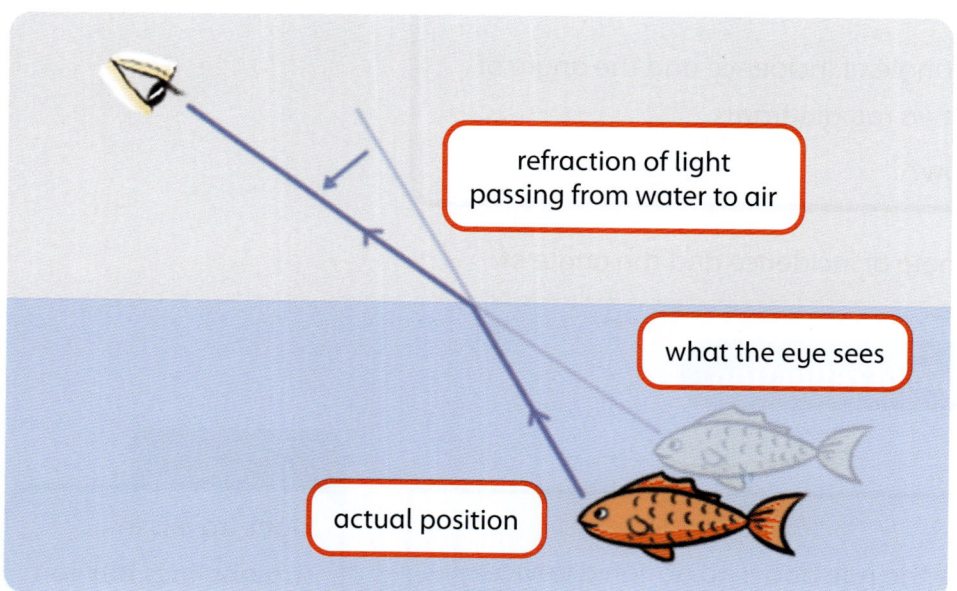

refraction of light passing from water to air

what the eye sees

actual position

Refraction makes the fish seem closer than it is

Talk about when you have seen things underwater. Describe to your partner what happens to the rays of light to explain what you saw.

■ For more activities, go to Workbook 6 page 84.

Magnificent colours

In a small group discuss these questions:

1 Name all the colours you have heard of.
2 What is your favourite colour? Why?
3 Why do people like to have flowers around?
4 Why are some places painted in bright colours?
5 What happens to colours as a room becomes darker?

Coloured discs

1 Take a circular piece of card and divide it into eight sections. Colour each section a different colour.
2 Push a stick or pencil through the centre of the circle.
3 Spin the stick.
4 What do you see?

Warning! Put the card disc on the desk before you push the stick through. Do not hold the disc in your hand. What could happen if you did?

When white light hits a surface, some of the colours are absorbed by the object and other colours are reflected. For example, a blue T-shirt absorbs most of the different colours and reflects only the blue colour.

If all of the light is reflected from a surface, the object will look white. If none of the light is reflected, the object will look black.

When you see a rainbow, it is because the rain droplets in the light are separating the white light into all the colours that make it.

We can use different coloured filters and your ray box to mix colours of light.

Can you explain why a leaf appears green?

List the colours of the rainbow. How did you remember them?

Science fact

White light is made up of all the colours of the rainbow. You can prove this by spinning a colour wheel.

Stretch zone

Why does a blue top look blue in daylight but look grey when it is getting dark?

Key ideas

- Light rays bend as they pass from one material to another.
- White light can be split into different colours.

■ For more activities, go to Workbook 6 page 85.

3 The Way We See Things

Can we see through it?

In this lesson you will explore how transparent materials let a lot of light through and opaque materials do not let light through.

Key words

opaque

translucent

transparent

Think back

Write down two properties of light.

List three opaque objects in the photographs.

List three transparent objects in the photographs.

An object that does not let light through is opaque.

Many objects are made of opaque materials such as fabric, wood, metal and ceramics.

Your clothes and large parts of cars, buses and trains are opaque.

An object that lets a lot of light through is transparent. Many objects are made from transparent materials. We need to look through windows and see through air.

Imagine if we could see through every material on Earth. What would life be like? Imagine if all the materials on Earth were opaque. What would life be like now?

■ For more activities, go to Workbook 6 page 86.

Some materials let a little light through. These materials are translucent. We can see shapes on the other side of translucent materials but not very clearly. Coloured and frosted glass are examples of translucent materials.

Look at the pictures. Discuss which of the materials is transparent and which is opaque. What can the person see through the translucent material?

 Grouping materials

You are going to investigate some materials to find out if they are opaque, transparent or translucent.

1   Your teacher will provide you with some different materials to test and some objects to look at.

2   Predict whether each type of material is opaque, transparent or translucent.

3   Look at each object through the different materials. Decide whether each material is opaque, transparent or translucent.

4   Make a table to record your results. To make your results reliable, repeat your investigation. Think about this when drawing your table.

5   Are there any items that did not give you the results you predicted?

Are there any links between the uses of the materials and whether or not they let light through?

Key idea

Transparent materials let light through and opaque materials block the light.

 Stretch zone

Research how glass is made translucent. Write down some examples of the uses of translucent glass you have seen.

■ For more activities, go to Workbook 6 page 87.

Making shadows

In this lesson you will observe how shadows are formed.

Key words
opaque
shadow

A shadow forms when light is blocked. Opaque materials will not let light through. This suggests that opaque materials form the best shadows.

Testing materials

You can investigate materials to test which make shadows. You need a light source and samples of different materials.

1 Set up your equipment.

2 Predict what you think will happen each time you try to cast a shadow using a transparent, translucent and opaque material.

3 Investigate which materials cast a shadow on the screen.

4 Record your results.

5 Compare your predictions with your results.

How accurate were your predictions?

Did opaque or transparent materials make the best shadows?

These are shadows of windows. Point to the part of the shadow that is the frame.

Point to the part that is the glass. Which one of these materials is opaque?

Does the photograph of the window shadows support this idea?

88

■ For more activities, go to Workbook 6 page 88.

Opaque materials can be very useful. We sit under sunshades or use umbrellas to keep the Sun off babies and young children.

Testing sunshades

Imagine you work for a sunshade company. Your job is to design a sunshade for the company to sell. You need to decide what type of material to use.

1. In your group, plan how you will set up this investigation.

2. Discuss your plan with the rest of the class before you start the investigation.

3. Use a light to represent the Sun. To make sure that this is a fair test, fix the light in position so it does not move and stays in the same place.

4. Hold each material in turn in front of the light and observe the shadow it casts.

5. Record your results. Conclude which material to use for the sunshade.

6. Compare your conclusion with the rest of the class.

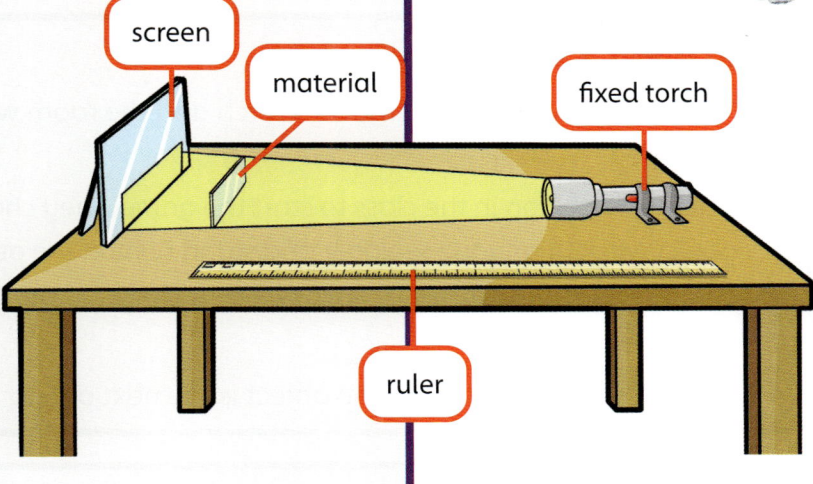

screen

material

fixed torch

ruler

Science fact

Eratosthenes, the head librarian of the Great Library of Alexandria, was the first person to calculate the size of the planet Earth. He used the size and angles of shadows. This was done over 2000 years ago.

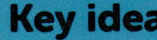

Stretch zone

Predict how far away you would have to hold a torch from an object that is 20 cm high to cast a shadow that is 60 cm high. Investigate to test your prediction.

Key idea

We can link the property of a material with how good it is at making shadows.

■ For more activities, go to Workbook 6 page 89.

Shadow games

In this lesson you will observe that shadows have the same shape as the opaque materials that block the light.

Key words
shadow
silhouette

Think back

Which type of materials allows a shadow to be cast? Write down three examples.

You can use shadows as part of games and challenges. To do that you must know about the ways shadows are made and cast.

Playing a shadow game

1 Your teacher will give everyone a torch and the room will be darkened.

2 Select one person in the class to start the game. They choose an object without anyone seeing it and stand behind the others.

3 The person then uses their torch to cast a shadow of the object onto a wall.

4 The first person to identify the object is the next one to play.

Be a scientist

When scientists use knowledge they have in a new situation, this is called applying knowledge. It allows them to solve problems or develop new ideas.

You are going to apply your knowledge of shadows to carry out some tasks.

▶ page 8

Making shadow puppets

You have learned a lot about shadows and light. Now you are going to write and perform your own short shadow puppet show.

1 Working in small groups, plan and write a short shadow puppet show to perform to the rest of the class.

2 Decide who the main characters are.

3 Using the materials provided, make puppets for your main characters. Think back to all the investigations you have carried out. This will help you to decide what size the puppets should be.

How can you make your puppet show more colourful?

4 Practise positioning the puppets in front of the light source.

5 In your group, perform your puppet show to the rest of the class.

You could also perform behind a sheet of fabric or a blind to make the images look different.

Why do shadow puppets have to be made from opaque materials?

■ For more activities, go to Workbook 6 page 90.

Can you make a coloured shadow?

1 Use a coloured filter to cover the front of a torch.

2 Test your filter by shining the torch against a wall. Record the colour you see.

3 Repeat this with three other filters.

4 For each filter use an opaque object to cast a shadow onto the wall. Predict the colour of the shadow that will be made.

5 Record the colour of the shadow for each filter.

6 Repeat the investigation but use a translucent object and then a transparent object to cast shadows. Predict the colours of the shadows you expect to see in each case.

7 Write a short report to explain your findings.

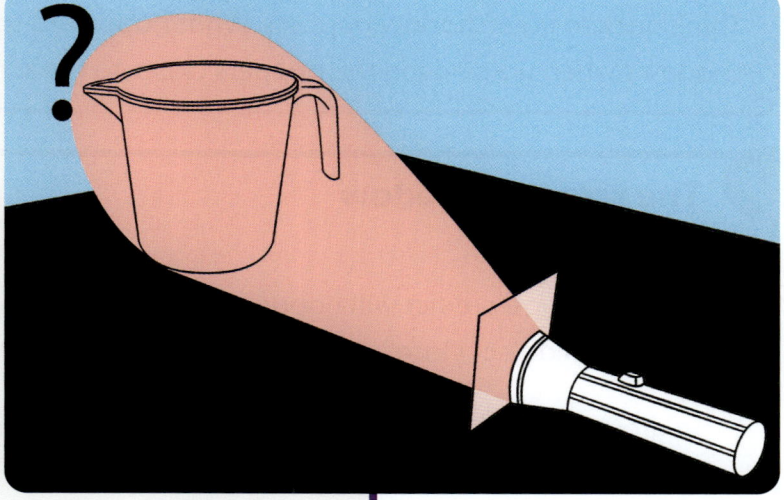

Student silhouettes

1 Shine a light at the side of your face. Ask a friend to tape some paper to the wall where your shadow appears.

2 Ask them to draw around your shadow on the paper and cut it out. You now have a silhouette of your head.

3 Make this into a shadow puppet and you can be in your play.

4 Display the silhouettes around the room. Can you guess who all of the silhouettes are?

Were some silhouettes more difficult to identify than others?

Stretch zone

Make an instruction leaflet to show other people how to make silhouettes. Include diagrams.

Key idea

Opaque materials can be used to make shadow shapes.

■ For more activities, go to Workbook 6 page 91.

Growing and shrinking shadows

In this lesson you will investigate how the size of a shadow is affected by the position of the object.

Key words

light source

opaque

shadow

Think back

Think back to your shadow puppets. What happened to the shadows when you moved the puppets closer to the screen?

The size of a shadow

What do you predict will happen to the shadow if we move an object closer to the light source?

1 In groups, carry out an investigation to explore if your prediction is true.

2 Using a torch as your light source, place it 2 m away from a screen or wall and fix it in place so it does not move.

3 Choose an opaque object that is easy to move. You will need to measure this object, so choose something that has a simple shape, such as a building block.

4 Place the object 10 cm in front of the light source and measure the length of its shadow.

What should you do next to make sure the results are reliable?

5 Now measure the size of the shadow from the same object at 20 cm, 30 cm, 40 cm and 50 cm away from the light source.

6 Record your results in a table like this one.

Distance from light source (cm)	Length of shadow (cm) Try 1	Length of shadow (cm) Try 2	Length of shadow (cm) Try 3

7 What do you notice about your results? Can you see a pattern?

8 Was your prediction at the start of the investigation correct?

Which variables were kept the same in the investigation to make sure it was a fair test?

■ For more activities, go to Workbook 6 page 92.

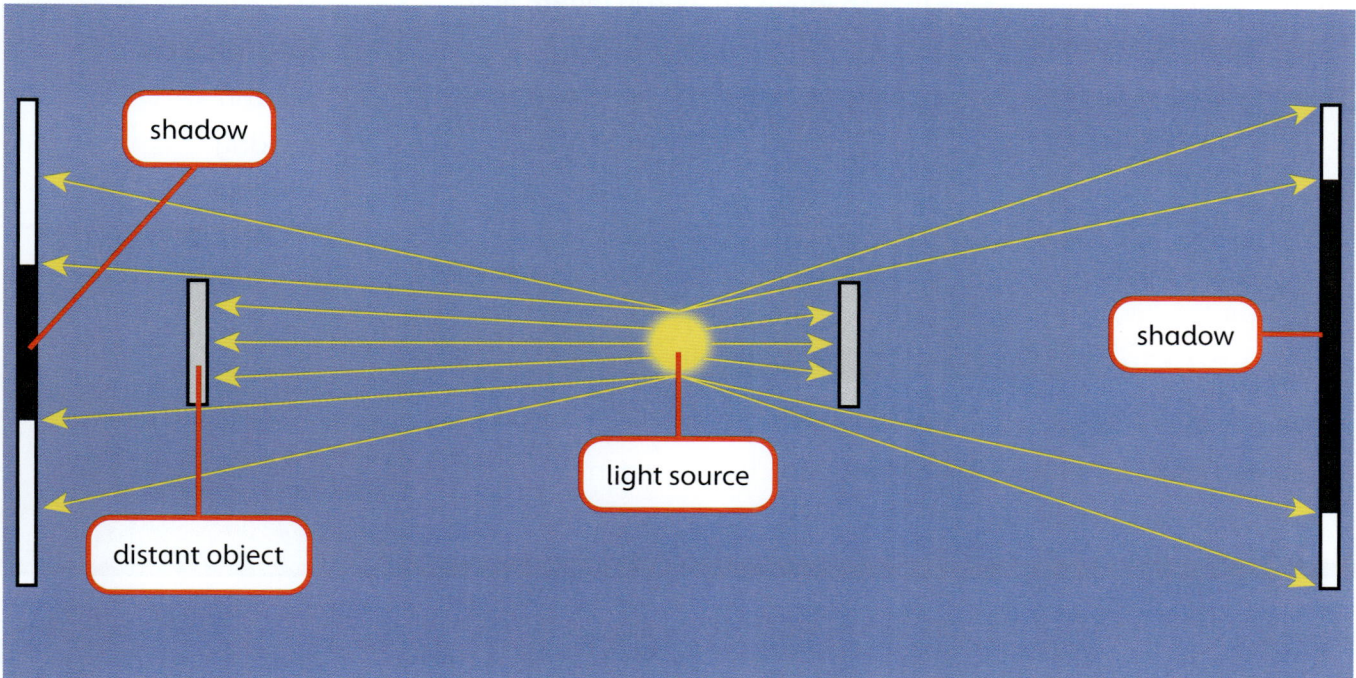

Look at the diagram above. When the light source moves closer to an opaque object the shadow gets bigger. If we move the light source away from an opaque object, the shadow gets smaller. This is because light beams travel in straight lines. They cannot pass through the opaque material and they cannot bend around it.

Moving shadow puppets

1 Design and make a puppet you can use to cast a shadow.

2 Use a torch to test if the shadow puppet can be seen clearly.

3 Explore how you can make your shadow puppet look like it is moving without actually having to touch it.

4 Investigate how you can make your shadow puppet look larger or smaller.

5 Present your ideas to the class. Show them your puppet moving and changing in size.

Stretch zone

Draw ray diagrams to show what happens to the shadow of an object as it is moved closer and further away from a light source.

Key idea

Shadows get bigger as the object is moved closer to the light source.

■ For more activities, go to Workbook 6 page 93.

Tracking moving shadows

In this lesson you will observe that shadows change in length and position throughout the day.

Key words

horizon

midday

shadow

Think back

What is the best natural source of light?

List two artificial sources of light that you have used in investigations so far.

During the day, the Sun appears to move across the sky. In the morning, the Sun is lower on the horizon and is not as powerful. At midday, the Sun is much higher in the sky and is at its most powerful. This makes the light seem brighter and it is hotter. After midday, the Sun gets lower towards the horizon again and is much less powerful and cooler than at midday.

Is the Sun really moving across the sky? What is actually happening?

The apparent movement of the Sun

1 Take your shadow puppet outside to make shadows or simply use yourself.

2 Change position during your investigation. Move your arms or legs. Try running on the spot or hopping.

3 Observe how your shadow moves. Does it move exactly like you? Draw around your shadow with chalk or mark it using pegs to show how it changes throughout the day. Record the time.

4 Keep a diary of your shadow. Check how long it is in the morning, at midday and in the afternoon.

5 Record your results in a table.

6 Observe how your shadow changes day to day and month to month.

It might not be possible to finish this investigation in one day but try to record your shadow every hour or as often as you can. Also, look at the shape of your shadow.

■ For more activities, go to Workbook 6 page 94.

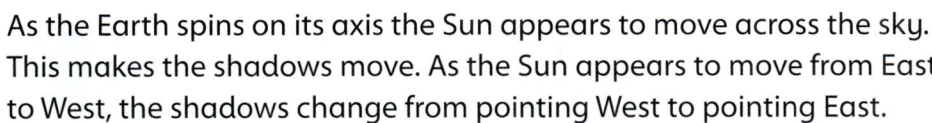

Does your shadow change shape?

Does your shadow change in size?

Does the darkness of your shadow stay the same?

7   Write a conclusion about your investigation.
    Use the following sentence starters to help you.

- During the morning my shadow is …
- At midday my shadow is …
- During the afternoon my shadow is …

As the Earth spins on its axis the Sun appears to move across the sky. This makes the shadows move. As the Sun appears to move from East to West, the shadows change from pointing West to pointing East.

The Sun also appears to move higher in the sky from dawn up until midday and then fall again until sunset. This makes shadows change from long to shorter and then long again.

Showing light and shadows

Artists have to show shadows in their pictures. It can make objects appear more realistic.

1   Choose an object and make a sketch.

2   Shade in the parts of the object that are in shadow.

3   Draw the object again but this time imagine the light is on the other side of the object.

4   Draw a larger picture with more than one object. Show the areas of light and shadow to make it look realistic.

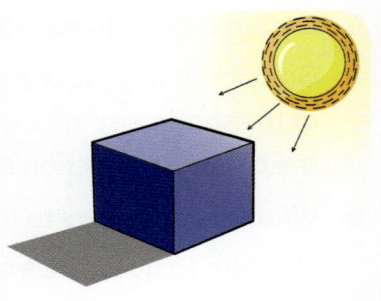

When planning buildings, architects have to think about how the Sun will appear to move and how this will alter the light and shade. The building may be in the shadow of another building at one part of the day, but in bright sunlight later in the day.

 Stretch zone

How do moving shadows impact on the light and shade where you live? Tell your partner how you could change things so you have more light in some places and more shade in others.

Key idea

Shadows change size and position throughout the day.

 For more activities, go to Workbook 6 page 95.

Shadow investigations

In this lesson you will investigate the changes in length and position of shadows throughout the day.

Key words

shadow

shadow stick

sundial

Think back

Do you know what this is?

What do you think it could be used for?

The shadow cast by the Sun will change throughout the day. Try this investigation. It builds on one you may have done a few years ago.

Shadow sticks

1. Find a place outside and push an upright stick into the ground.

2. Use a compass to mark on the ground where north, south, east and west are.

3. Draw around the shadow of the stick at the start of your investigation. You can do this on paper or use chalk on the ground. Observe the compass direction that the shadow is pointing in. If it is between two, then you can show this as N/E or S/W.

4. Design a table to record your results.

5. Go back to your stick every hour and mark the new position of the shadow so that you can measure its length. Record the compass direction it is pointing towards.

6. Write your answers to the following questions in your notebook.

   - What do you notice about the length of the shadow?

   - What do you notice about the direction of the shadow?

7. You could repeat this experiment for a day every month.
   What do you predict will happen to the shadows?

8. Write down two ways that you could make your investigation more accurate.

9. Use your results to find out the exact time of midday without looking at a watch or clock.

It is possible to find out the exact time of midday with the results from this investigation. When the shadow is at its shortest, the Sun is directly above you. This is midday.

■ For more activities, go to Workbook 6 page 96.

Tracking your shadow

1 Work in groups of three or four and find a clean, smooth and safe surface outside.

2 Draw a mark on the ground so that you stand in the same position every time. You could draw around your feet.

3 Stay in one position and stand very still. The other people in your group should use chalk to draw around your shadow on the ground. (You may prefer to use paper.) Make sure nobody erases the diagram of your shadow.

4 If possible, revisit the drawing of your shadow every hour and draw around your new shadow. Over the course of the day, your shadow should change in size and position.

5 Try to check if this changes over a week.

What do you predict will happen to the length of your shadow over a month?

Look at the picture. Decide with a partner where you would place a sticky note to label the shadow drawn at midday. Share your reasons with the class.

These investigations show that it is the apparent movement of the Sun across and up and down in the sky that change the length and direction of shadows. A shadow will always be pointing in the opposite direction of the light source. This means that shadows cast by the Sun in the morning, when the Sun is in the east, will point west.

Stretch zone

Plan a way that you could use the time of day and shadows to find out which direction is north, south, east and west. Demonstrate your method to a partner.

Key idea

We can use the size and direction of a shadow to tell the time of day.

■ For more activities, go to Workbook 6 page 97.

Light intensity

In this lesson you will find out that light intensity can be measured.

Key word
light intensity

Think back

Think about the ray box you made earlier.

We can describe how bright or intense a light source is by using words such as very bright, dim, dark or glowing. It might even be possible to compare different light sources. We know that the Sun is brighter than a candle and a lighthouse is brighter than a torch. However, this still isn't very scientific. Scientists need to measure light intensity much more accurately.

Light intensity is how much light energy there is in one place.

Directing the light through the tiny slit in a ray box makes the light seem brighter or more intense. The wider the slit, the less intense the light.

Brightness and light intensity are similar, but what appears bright to one person may not seem as bright to another person. We need a way of measuring light intensity.

Why is it better to have a number to measure light intensity rather than using words? Think about other measurements that we use in science.

A photodiode

A photodiode changes the light energy into an electrical current. This is displayed as a number. These devices are sometimes called light meters.

Light meters are used to measure the strength or brightness of electric light bulbs. Electricians also use them to measure the light intensity in a room. Some rooms need to have more light intensity than others.

Hospital operating theatres need a very high light intensity. A bedroom does not need a high light intensity.

Why does an operating theatre need a high light intensity? Explain your answer to your partner.

■ For more activities, go to Workbook 6 page 98.

Light meters are also used in digital cameras and mobile phone cameras. They measure the light intensity around an object. If there is not a high enough reading, the flash is used.

How do our eyes change the amount of light that enters?

Why can't we see very well in a low light intensity?

Have you ever been on a road or in a big building and the lights have come on automatically? This is because a light meter has recognised that there isn't enough light.

Look carefully at a solar calculator and notice the solar panel strip. This transfers light energy to electrical energy to enable us to use the calculator.

Measuring light intensity

We can measure how much light a calculator needs to work.

1 Place the calculator beneath a bright light source. A small table lamp or torch fixed in place will work well.

2 Switch the calculator on and enter the number 1000 to check that it is working.

3 Cover the solar panel with one piece of tissue paper. Check to see that the number is still showing.

4 Record this in a table of results in your notebook.

5 Keep adding layers of tissue paper one at a time and filling in your results table.

6 Record the number of layers of tissue paper needed to stop the calculator from working.

What do you think would happen if you covered the solar strip?

How can the calculator and tissue paper be used to measure the light intensity of different light sources?

Stretch zone

Write a plan for an investigation to compare the light intensity of a torch, a candle, the Sun and a table lamp. Ask your partner to suggest improvements.

Key idea

Light sources vary in brightness. It is important that we can measure this accurately.

3 The Way We See Things

■ For more activities, go to Workbook 6 page 99.

Using scientific methods to measure light intensity

In this lesson you will find out how light intensity can be measured.

Key words

light intensity

lumen

lux

What might happen if we had no way of measuring the brightness of light sources?

Before the 18th century, not many people measured the intensity of light. Some astronomers tried to use light levels to estimate how far away planets were from the Earth.

At this time, light intensity was measured in foot-candles. This is a very old English method of measuring light intensity.

During the 19th century, the electric light was invented. People tried to prove that the electric light was more efficient than the gas light. In 1920, light intensity was measured in an accurate way by using luminance. Luminance is the total amount of visible light present and is measured in lumen.

Light intensity is also measured in a unit called lux. 1 lux is equal to 1 lumen.

When building offices and factories, there are strict rules about light intensity. If the light is too bright, workers might suffer from headaches and the cost of electricity will be high. If the buildings are too dark, they can be dangerous and workers may not be able to see to work properly.

There are fewer rules when it comes to our own homes. It is still important to make sure there is enough light, though, to see clearly.

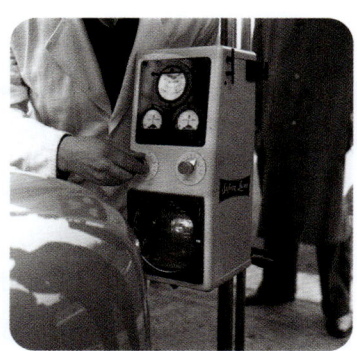

Measuring the light intensity of a car headlight

Activity	Foot-candles	Lux
washing clothes	10–15	100–150
eating	20–30	200–300
reading/studying	50–150	500–1500
sewing	100–200	1000–2000

Light intensity levels needed for activities at home

Look at the results in the table. Which two activities need the most light? Why?

■ For more activities, go to Workbook 6 page 100.

Presenting light intensity data

If you have ever stood in a darkened room with just one source of light, have you noticed how the light fades as you move away from the light?

The table below shows some results from such an investigation.

Distance from source (cm)	Light intensity (lux)
10	930
15	410
20	235
25	155
30	100
50	40
70	20
80	15
100	10

Distance and light intensity

Be a scientist

When scientists use data from another source it is called secondary data. They always check that the data has come from a reliable source and has not just been made up.

▶ page 8

1 Draw a graph in your notebook to show these results.

2 Compare your graph with others in your class. What is similar and different about them?

3 Use the patterns you can see in the data to reach a conclusion.

The table and your graphs should show that the further away from a light source you are, the less bright the light will be.

Stretch zone

Research secondary sources such as books and the internet to find out some of the brightness of common stars near our solar system. Make a table to present your results.

Check how much you know.
Try the questions on pages 102–103.

Key idea

Scientists have developed accurate ways to measure light intensity.

■ For more activities, go to Workbook 6 page 101.

What have I learned about the way we see things?

1 Look at the pictures below.

a   Which picture shows a translucent material?   _____

b   Which picture shows a transparent material?   _____

c   Which picture shows an opaque material?   _____

2 Tick the two correct statements below.

Shadows are at their shortest in the morning and evening. ☐

Shadows always point in the opposite direction to the light source. ☐

Shadows are at their shortest at midday. ☐

Shadows are the same length throughout the day. ☐

3 Circle the correct answer.

a   The bending of light as it passes from one material to another is called:

reflection   radiation   refraction

b   The bouncing back of light from a surface is called:

reflection   radiation   refraction

4 Draw the shadow that would be made by the vase on the picture.

■ For more activities, go to Workbook 6 page 102.

5 Label the diagram. Use the words in the word box.

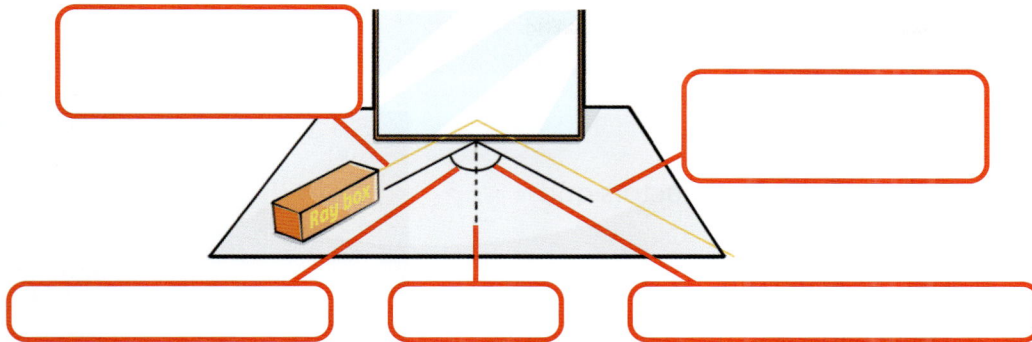

angle of incidence
angle of reflection
incident ray
normal
reflected ray

6 a Label the diagram. Use the words in the word box. One word can be used more than once.

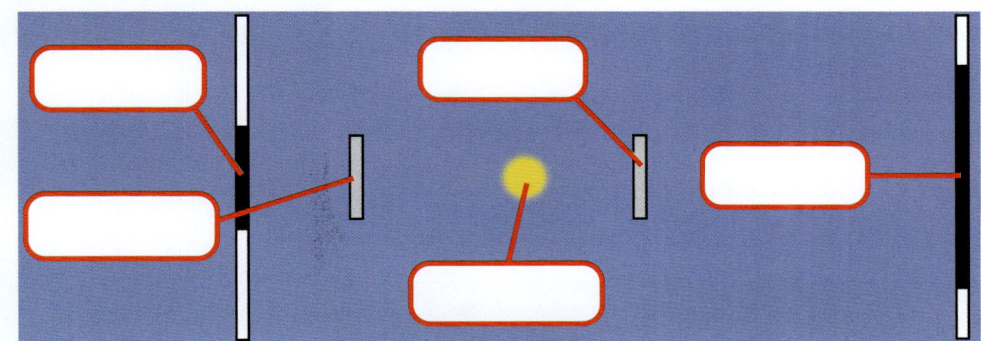

distant object
light source
near object
shadow

b Draw in the rays of light that leave the light source until they reach the screens.

c Draw in the shadow on the right-hand screen. Draw it the correct size.

7 The unit used to measure light intensity is the lux. Study the table below.

Distance from source (cm)	Light intensity (lux)
10	930
15	
20	235
25	155
30	
50	40
70	
80	15

a Insert the missing values for light intensity in the correct place in the table.

20   100   410

b Analyse the results to allow you to tick the correct statement below.

Light intensity stays the same no matter how far away the light source is. ☐

Light intensity increases the further away from the light source we are. ☐

Light intensity decreases the further away from the light source we are. ☐

c Predict the light intensity if the distance from the source was 12.5 centimetres. _____

■ For more activities, go to Workbook 6 page 103.

4 Building Electrical Circuits

In this unit you will:

- find out how some materials are better conductors and insulators of electricity than others
- understand why metals are used for cables and wires and why plastics are used to cover wires and as covers for plugs and switches
- learn how changing the number and voltage of cells affects the components in a circuit
- predict and test the effects of making changes to circuits
- draw diagrams of series and parallel circuits using standard symbols.

When you comb your hair, or rub a balloon on your clothes, a small amount of electricity is made.

Why does the balloon pick up the pieces of paper?

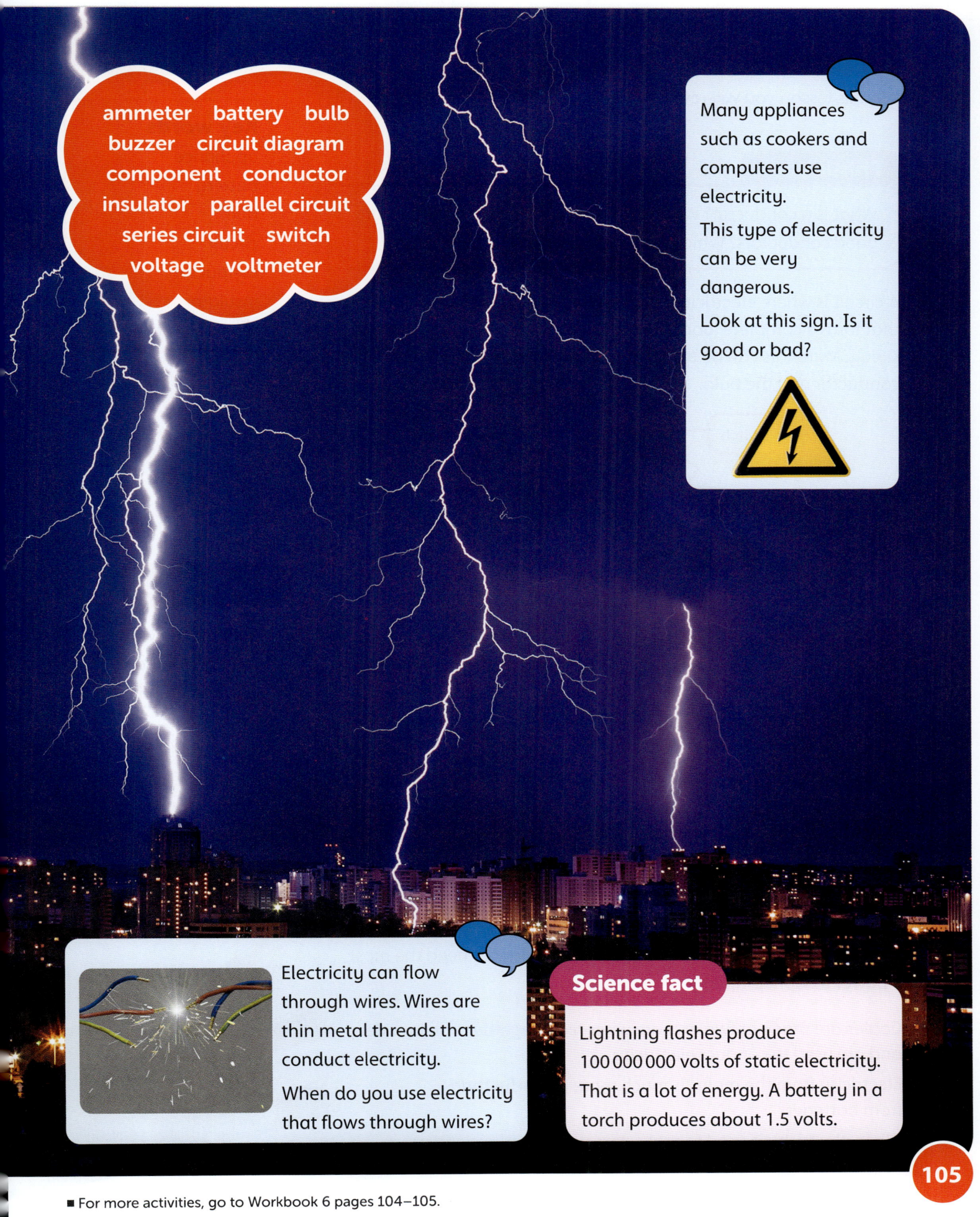

ammeter   battery   bulb
buzzer   circuit diagram
component   conductor
insulator   parallel circuit
series circuit   switch
voltage   voltmeter

Many appliances such as cookers and computers use electricity.

This type of electricity can be very dangerous.

Look at this sign. Is it good or bad?

Electricity can flow through wires. Wires are thin metal threads that conduct electricity.

When do you use electricity that flows through wires?

Science fact

Lightning flashes produce 100 000 000 volts of static electricity. That is a lot of energy. A battery in a torch produces about 1.5 volts.

■ For more activities, go to Workbook 6 pages 104–105.

Revising electricity

In this lesson you will revise the basics of electricity.

Key words

appliance
battery
bulb
circuit
conductor
insulator
mains electricity

Think back

What is a conductor? What is an insulator? Which materials are good conductors? Which materials are good insulators?

A circuit is where electricity can flow through a series of conductors that form a complete loop. The conductors are labelled in this test circuit. Materials can be tested by completing the circuit between the conductor and the bulb.

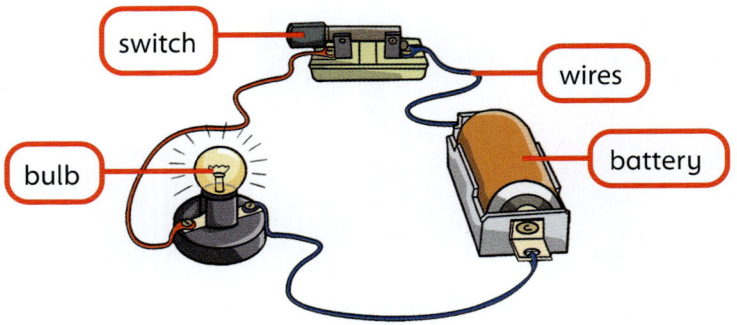

If the bulb lights up brightly, then lots of electricity is passing through the material from the battery to the lamp or bulb. This material must be a good conductor.

If the bulb does not light up, the material must not allow electricity to pass through it. It is an insulator.

Some electrical appliances use batteries as a source of their electricity.

In many cities and homes we use electricity that flows from a power station through cables. This is called mains electricity.

With a partner, agree three appliances that use batteries. Now agree three appliances that you plug into the mains electricity. Look around the room for clues. Which appliances need the most power to work properly?

Cables are made of two or more wires fixed together side by side and are used to conduct electricity.

Electricity from the mains is much more powerful than from batteries.

When electricity is used to make something work, scientists say electricity powers it.

Science fact

Humans are very good conductors of electricity! If electricity is conducted through humans, they may get an electric shock. This can be very serious. Electricity can kill!

106

■ For more activities, go to Workbook 6 page 106.

> Can you use batteries to power an electric oven? Explain your answer.

Electricity flows through a circuit when everything is joined up correctly. We call this electricity the current. The current flows from the power source, which produces the electricity, through conductors, and back to the power source. This is why it is called a circuit, like a running track.

If there is a break in the circuit, like in the test circuit shown here, the electricity will not flow. The electric oven and your other appliances will not work.

Test circuit

That is why good conductors are used in homes, schools and hospitals to make sure all the electrical appliances work when they are needed.

Insulators are useful because they stop electricity from flowing into us. They can stop or slow down the flow of electricity because they do not conduct electricity.

Is it an insulator?

1   Set up a test circuit using a battery, a bulb, wires and connectors as shown above. Leave two connectors free so that you can test materials.

2   Test your circuit to make sure everything is working. Choose a material that you know is a good conductor.

> If everything is working, what will happen to the bulb?

3   When your circuit is working, predict which materials will be a good insulator.

4   Test the materials and record your findings.

> Which material would you use to insulate a wire?

Key ideas

- Mains electricity and batteries are used to power appliances.
- Some materials are better conductors of electricity than others.

■ For more activities, go to Workbook 6 page 107.

Choose your conductor

In this lesson you will find out which materials are good conductors of electricity.

Key words
ammeter
conductor
current
insulator
metal

Think back

Remember from your investigations which materials are good conductors. They are usually metals.

Discuss any metals you can think of. Choose two and discuss what they are used for.

Electric tools like drills have to have some parts that let electricity flow through them and other parts that do not.

Science fact

The first hand-held electric drills were invented in 1895. We now have portable drills that do not need cables. They are powered by batteries.

Discuss the advantages and disadvantages of battery-powered drills over drills powered from mains electricity.

There are some non-metals that are good electrical conductors too. For example, the graphite in your pencil and also tap water and sea water are good conductors. That is why you must never spill water near an electrical socket as it can conduct electricity into you!

You now know about the different materials that are good conductors. They are mainly metals. Now we will find out if some metals are better conductors than others.

Warning! Never put electrical appliances near water. It is very dangerous. Water is a good conductor of electricity.

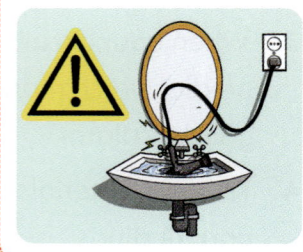

Stretch zone

Why do you think humans are good conductors of electricity?

■ For more activities, go to Workbook 6 page 108.

Electrical conductivity is the term used to explain how a material lets electricity through it.

Which metal is the best conductor? Part 1

1 Set up a test circuit like the one on page 107. You are going to investigate the conductivity of different metals.

2 Copy and complete the table below to record your observations. One example is given to you to get started.

Metal	How bright is the bulb?
copper	very bright

Remember: the brighter the bulb, the better the conductor is.

Be a scientist

It is a science skill to be able to criticise your work and come up with better ways to test things. This is called evaluating.

▶ page 13

3 Look at your observations. Are some metals better conductors than others?

4 Write down some ways you can improve your investigation to make it a fair test.

In this investigation, you used the brightness of the bulb to measure the amount of current flowing through the circuit. Scientists use a piece of equipment to measure the current in a circuit. It is called an ammeter. The stronger the current is, the higher the number of amperes (or amps) shown on the ammeter.

Which metal is the best conductor? Part 2

In your group you will again investigate which materials conduct electricity the best. This time use an ammeter to measure which material allows the most current through.

1 Make a table in your notebook to record your results.

2 Order your results to evaluate which material is the best conductor and which is the worst conductor.

Did you find that metals were good conductors?

Copper, steel, gold, silver and iron are some of the best electrical conductors.

■ For more activities, go to Workbook 6 page 109.

Science fact

A French scientist called André-Marie Ampère invented the ammeter.

The unit used to measure current is the ampere (or amp). The amp is named after him. The symbol for the ammeter is

An ammeter measures the current flowing through a circuit.

Key idea

Some metals, graphite and water are good conductors of electricity. Most other materials are poor conductors.

Using metals and plastics in electrical circuits

In this lesson you will learn why metals are used for cables and wires and why plastics are used to cover wires and as covers for plugs and switches.

Key words

cable

metal

plastic

plug

wire

Not all plugs are the same but they do the same job

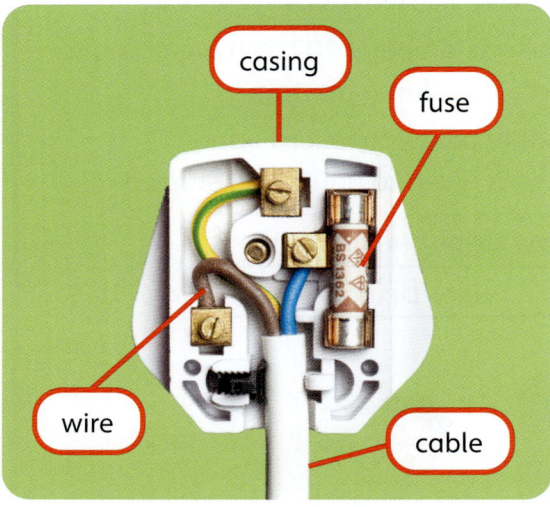

casing

fuse

wire

cable

Look at the photograph of the inside of a plug. Make a list of all the materials you can see.

Copper is usually used in the wires in your classroom and at home, and in electrical sockets. This is because copper is a good conductor and it is not as expensive as some other metals such as gold. It also lasts a long time.

Metals used to make wires must have other properties, not only conductivity. They need to be able to be stretched and pulled into wires.

Analysing data about metals

Look at the information in the table below. Then answer the questions in your notebook. Explain your answers fully.

1 Why is mercury not used in wires?

2 If you could not use copper, which metal from the list would you use?

3 Why do you think gold is not used in wires often?

4 Why is copper used in wires more than other metals?

Metal	Conductivity	Can be pulled into a wire	Cost
gold	very good	very easy	very expensive
copper	excellent	very easy	very cheap
mercury	good	not possible	expensive
graphite	OK	not possible	cheap
aluminium	very good	very easy	cheap

■ For more activities, go to Workbook 6 page 110.

Overhead cables sometimes have to stretch across very large distances. They connect homes and other buildings to the main supply of electricity. Scientists and engineers work together to find the best materials and methods to make things work.

Aluminium is the best metal for overhead cables. This is because it is cheaper than other metals. It is also strong when it is made into wires.

The aluminium cables carry more than 765 000 volts of electricity. The batteries you used in your investigations carry about 1.5 volts.

Science fact

The longest overhead cable is 2.7 kilometres. It connects Zhoushan island to the Chinese mainland.

Why are overhead cables from power stations made from aluminium?

Uses of materials

Look at the photograph of the inside of a plug opposite. There is a casing and the wires are covered in different colour materials.

1 What material is the casing of the plug made from?

2 Why is this material used?

3 What can happen if the insulation around a wire breaks?

Key ideas

- Metals are good conductors. They are used for cables and wires.
- Plastics are good insulators. They are used to cover wires and they are also used as covers for plugs and switches.

4 Building Electrical Circuits

111

■ For more activities, go to Workbook 6 page 111.

Changing circuits

In this lesson you will predict and test the effects of making changes to circuits.

Key words

circuit

component

Think back

Think back to your work in previous years to predict what will happen to the circuit if:

- you add more bulbs
- you increase the number of batteries
- you increase the length of the wires.

Be a scientist

Remember a prediction is not a guess. It is based on ideas or results that you have learned about before.

▶ page 8

Changing parts of a circuit: Part 1 – planning

1. In your pair, plan an investigation into what happens when you change parts of your circuit. Start by making a simple circuit like the one here and then decide what you are going to change.

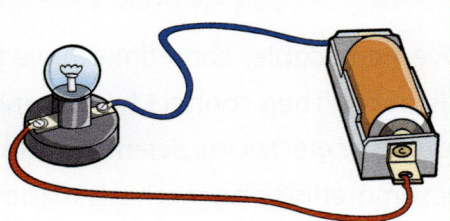

2. Write down the answers to these questions in your notebook to help you think about your plan.
   a. Are you going to change the number of bulbs?
   b. Are you going to change the number of batteries?
   c. Are you going to change the length or thickness of the wire?
   d. What are you going to measure?

   For example, you might decide to investigate the thickness of the wire. Your teacher will give you wires of different thicknesses. You can predict which ones will make the bulb brighter.

3. To make the test fair, you must only change one thing at a time. For example, if you are investigating the length of the wire, then the number of batteries and bulbs must stay the same.

What will you keep the same in your investigation to make it a fair test?

4. In your notebook, draw a picture of the circuit you are using to test what happens when you make changes to your circuit.

How will you record your results?

■ For more activities, go to Workbook 6 page 112.

Changing parts of a circuit: Part 2 – carrying it out

1   Now carry out your investigation.

2   Record the changes you made.

3   Write down what happened to your circuit.

4   Write a conclusion from your results. Compare your results with other groups in your class.

 Be a scientist

Scientists discuss their conclusions with others to help to make sure their arguments are sensible.

▶ page 12

How did you make sure you carried out a fair test? What did you do?

Batteries, or cells, push electricity around a circuit. The bigger the voltage of the batteries, the bigger the push. A bigger push gives a bigger current and bulbs in the circuit will shine brighter.

Adding more bulbs in a line to a circuit (in series) means the bulbs share the voltage so each gets less. Each bulb will be dimmer than if it was the only bulb in the circuit.

 Stretch zone

Research what happens if too much voltage flows through a component such as a bulb. How can you prevent this from happening? Share your ideas with a partner.

Key idea

You can predict what will happen before you make a change to a circuit. Then you can make the change to test your prediction.

■ For more activities, go to Workbook 6 page 113.

Circuit breakers

In this lesson you will explore how components work when there are breaks or switches in the circuit.

Key words

circuit breaker

fuse

gauge

switch

Think back

Can you describe what a switch is used for?

What will happen if we do not have switches in our homes or school?

Switches are used in electrical circuits to stop the flow of electricity through the components. It is a controlled break in the circuit. When you turn on the switch, the circuit is complete and the electricity flows through the switch to the component.

When the switch is turned off the electricity does not flow. It stops at the opening between the metal parts inside the switch.

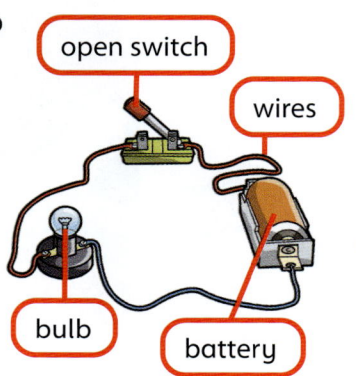

open switch

wires

bulb

battery

What will happen to the bulb in a circuit with a closed switch?

What about with an open switch?

If we did not use switches, the electricity would be either on or off all the time. If it was flowing all the time, it would be bad for the environment, expensive and very dangerous. Components can get very hot if they are left on for long periods of time. This can cause fires.

Science fact

Switches that switch off the electric current automatically if something goes wrong are called circuit breakers. Small ones are found in houses and large ones can protect cities.

Warning! Be careful touching an appliance when it is switched on. Discuss why this is important.

The first photograph shows an electrical safety component. It is called a fuse. It works in the same way as an old-fashioned bulb, like the one in the second photograph.

Look closely at the bulb and the fuse. Can you see any similarities?

■ For more activities, go to Workbook 6 page 114.

When the electricity flows through the wire in a bulb, it increases in temperature. Inside the bulb there are gases which encourage the wire to glow. The shape of the glass bulb makes the light glow. This is how a bulb gives us light.

Does the width of a wire affect how hot it gets?

Your teacher will give you different gauges or widths of the same kind of wire to investigate. You are going to observe how the wires glow in the same circuit.

Warning! Do not touch the wires as they could get very hot.

1. Make a simple circuit to test your wires. Include wires, a bulb, a battery and connectors. In this investigation, the bulb is included so that you can see if the circuit is working.

2. Plan your investigation.

3. Design a table to help you record your observations.

4. Connect a length of the wire into the circuit. Leave everything connected for 30 seconds. Make sure the bulb is glowing.

5. Can you see if the wire is beginning to get hot? Is it glowing? Record your observations.

6. Repeat this for at least three different widths.

Why do the wire samples have to be the same kind?

Does the length of the wire have to be the same?

Did any of the wires snap or break?

Be a scientist

Scientists have tested the gauge of wires to see how long they can allow current to flow without breaking. They have used this evidence to develop safe appliances.

▶ page 8

The type and gauge of wire is designed especially for the bulb. It gets hot enough to give out light, but if it gets dangerously hot, it breaks.

A fuse works in the same way. The wire in the fuse is designed to break if too much electricity flows through it. This is a safety device known as a circuit breaker. It prevents too much electricity flowing into appliances and breaking them or causing electrical fires.

Stretch zone

Research some other circuit breakers. Write a short report on how and when they are used.

Key idea

Fuses and bulbs can act as circuit breakers in electrical circuits.

■ For more activities, go to Workbook 6 page 115.

Using circuit diagrams

In this lesson you will draw diagrams of series circuits using symbols.

Key words

cell

circuit diagram

circuit symbol

series circuit

Think back

List the components of electrical circuits you have learned about. What does each one do?

A series circuit is an electric circuit which has only one path through it.

Scientists from all over the world use the same symbols to help them share ideas about circuits. New ones are being invented all the time.

It takes time to draw the different types of circuits. Think how long it took you to draw a circuit. To save time, we use symbols to draw diagrams of circuits. All the things you drew are called components, or parts. There is a symbol for every component we use.

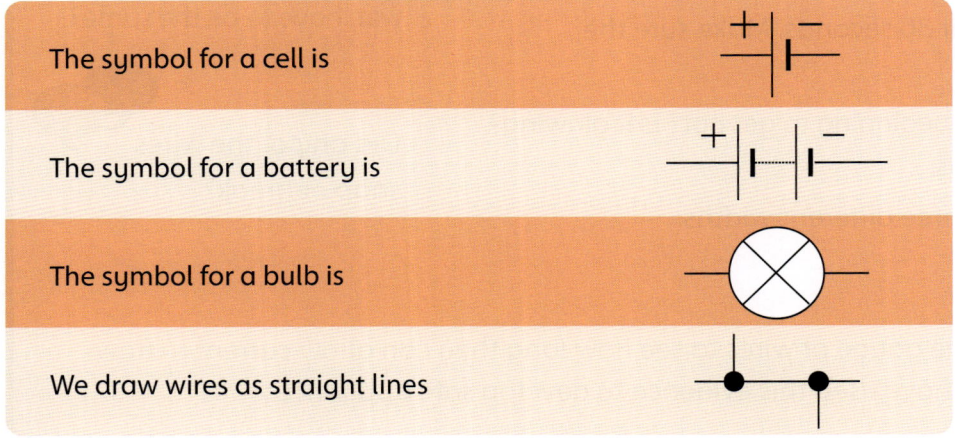

The symbol for a cell is	
The symbol for a battery is	
The symbol for a bulb is	
We draw wires as straight lines	

What is a battery? How is it different to a cell?

A battery is a collection of smaller units called cells. One end of a battery is positive (+). The other end is negative (−).

When you make a circuit with more than one battery you must set up the batteries correctly. You must join together opposite ends of the battery so that the current flows round the circuit. You must not join two positive or two negative ends together. If you join the same ends together the current will not flow round the circuit.

Why does the circuit have to be complete for electricity to flow?

Stretch zone

Some scientists say that a battery acts like a water pump. Can you use the information above to explain this idea?

116

There are many important symbols that you need to learn. You know the symbols for cell, battery, bulb and wire.

Look at the symbols below.

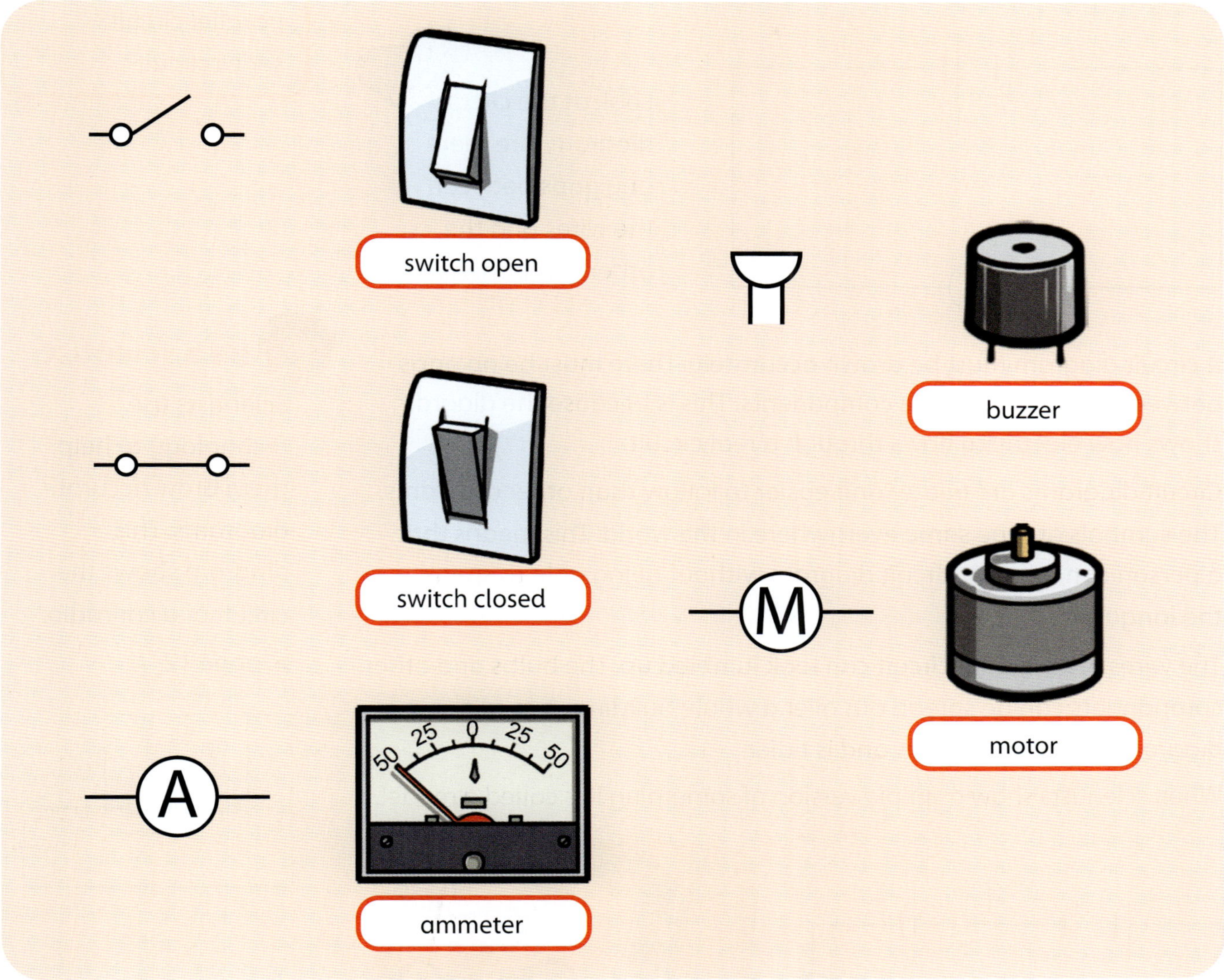

switch open

switch closed

ammeter

buzzer

M

motor

You need to practise drawing circuits with these symbols.

Drawing circuit diagrams

1 Draw the following circuits in your notebook:

   a two bulbs, one battery, wires, an open switch and a motor

   b two batteries, wires, one bulb, a closed switch and a buzzer.

2 Will both of these circuits work?

3 Build the circuits to support your answer to part 2.

Key ideas

- You can draw series circuits.
- Symbols show different parts of a circuit.

■ For more activities, go to Workbook 6 page 117.

Types of circuits

In this lesson you will compare and draw series and parallel circuits using symbols.

Key words

circuit diagram
circuit symbol
parallel circuit
series circuit

Electricity cannot cross gaps. It has to travel through a conductor.

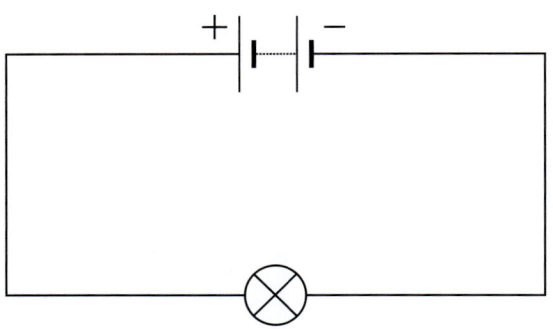

Look at the circuit diagram. Are there any gaps between the wires and the other components?

A circuit diagram must be drawn accurately. There must be no gaps between the wires and the components. This is because the diagram shows how electricity flows through a real circuit.

Circuit diagrams are neater and easier to follow than other diagrams. The symbols are the same everywhere in the world. This means that a person can follow a circuit diagram without being able to speak the language.

The wires in an actual circuit can look jumbled up. The bulbs and other components can be spread out over a whole building. A circuit diagram uses straight lines and is tidier.

When scientists show a simple version of something it is called a model.

**Be a scientist**

Scientists use technology to help them draw accurate diagrams. This diagram shows the circuit of a car radio.

▶ page 12

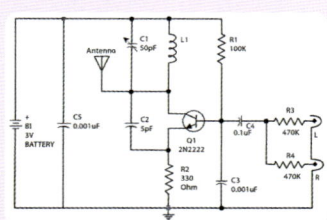

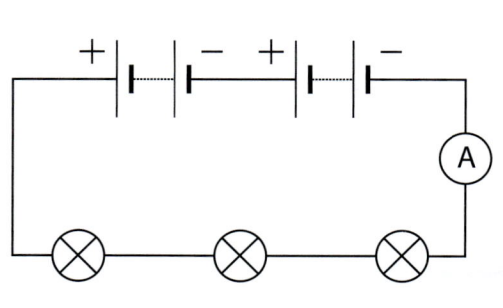

Look at the diagrams of a circuit. Discuss these questions.

- Which of the diagrams shows a circuit diagram?
- Which of the diagrams is easier to understand?
- Why do you think circuit diagrams are used worldwide?

118

Parallel circuits

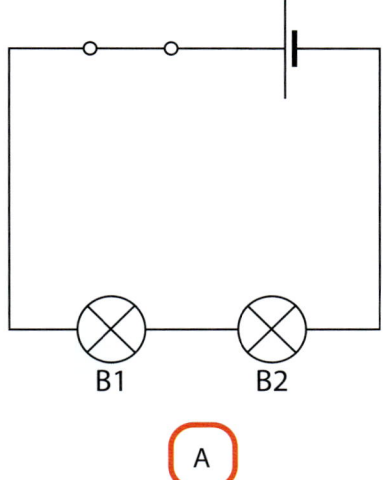

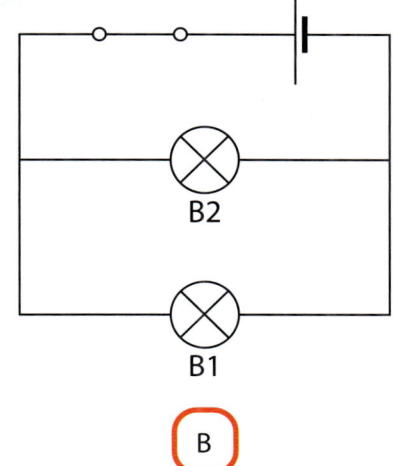

Discuss the circuits. Name all the components. What will happen to bulb B1 if the bulb B2 is replaced with connecting wire in each circuit?

There are different ways of arranging electrical circuits. You have already seen one way of arranging a circuit. This is shown in circuit A. The bulbs follow each other. This is a series circuit.

Look at circuit B. The bulbs do not follow each other. They have their own branch of the circuit. This is a parallel circuit.

Comparing series and parallel circuits

1 Set up the circuits shown in diagrams A and B but do not close the switches.

2 Before closing the switches, predict which circuit will have the brightest bulbs.

3 Close the switches and compare the brightness of the bulbs. Record your observations. Was your prediction correct?

4 Remove bulb B2 from each circuit and close the circuit. Record your observations. Was your prediction in your discussion correct?

5 Produce a short information sheet to tell people about the similarities and differences of series and parallel circuits.

Stretch zone

Research the circuit for a set of traffic lights or a burglar alarm. Try to download a circuit diagram to add to your notes.

Science fact

Parallel circuits are used in houses, streetlights and many decorative lights. This is because each bulb is brighter than in a series, and if one bulb fails, the others stay on.

Key idea

Components in a circuit can be set up in series or in parallel.

4 Building Electrical Circuits

119

■ For more activities, go to Workbook 6 page 119.

Measuring voltage

In this lesson you will investigate how the brightness of a lamp or the volume of a buzzer changes with the voltage in a circuit.

Key words

component
parallel circuit
voltage
voltmeter

Think back

Think back to the circuit diagrams you have used. Why do we call these simple series circuits?

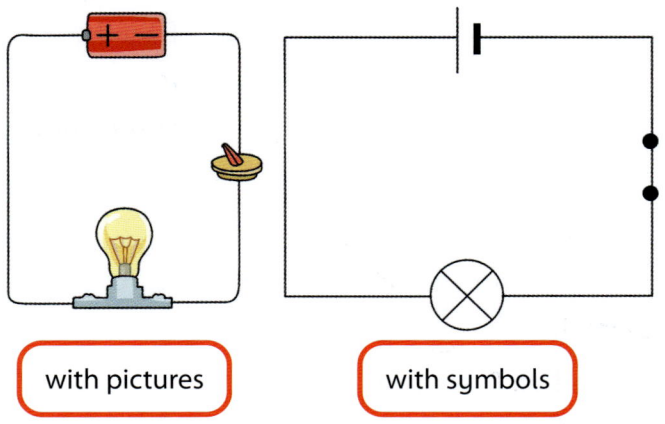

with pictures

with symbols

List the components in the circuit. Would the bulb be lit or not?

Think back to investigations you have carried out. When you add more batteries to a circuit, the bulb gets brighter.

Talk to your partner about why you think this happens.

Voltage describes the electrical force needed to push an electric current between two points. It can be thought of as the pressure from the power source that pushes the current around the circuit. In the above circuit, the power source is the battery.

We can say that voltage equals the pressure in a circuit. It is measured in volts with the symbol V. The voltmeter is very similar to an ammeter, which measures current. There is always a V displayed to help us.

A voltmeter measures the electrical difference between two points in a circuit.

A voltmeter has to be connected in an additional circuit to the one where the measurement is made. This is called a parallel circuit. This is because a voltmeter can affect the flow of electricity in a series circuit.

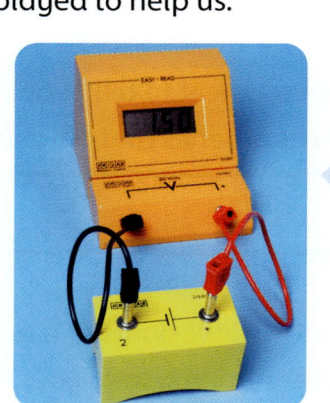

Look at this voltmeter. What reading is displayed?

■ For more activities, go to Workbook 6 page 120.

If the voltmeter was connected in series in the circuit, it would not record the correct measurement.

Using a voltmeter

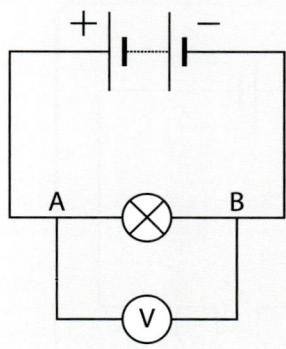

1 Set up your circuit using the diagram.

2 If there isn't a reading displayed on the voltmeter, what should you do? Think back to other circuits you have investigated. How can you test the components in the circuit?

3 Sometimes the reading shows a negative symbol. Change the wires to the voltmeter around to make sure the current is flowing correctly.

What could the negative reading tell you about the flow of the current in the circuit?

4 Move the voltmeter to different places in the series and parallel circuit. Record the readings from the voltmeter.

5 Draw a circuit diagram of each position and record the voltage readings on the diagram showing the position of the voltmeter.

6 Compare the readings from the series and parallel circuits. Write a concluding sentence about how the position of the voltmeter affects the voltage reading in a circuit.

Science fact

The electric eel, a type of fish, can deliver a shock of up to 600 volts when hunting or for self-defence.

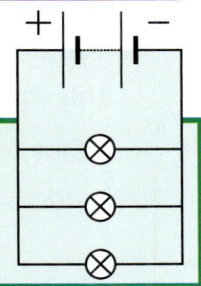

Key ideas

- Voltage is the difference of the force of electricity between two points in a circuit.
- A voltmeter measures voltage. It must be used in a parallel circuit.

Stretch zone

Study the circuit. Explain where you should place a voltmeter to measure voltage around the circuit.

■ For more activities, go to Workbook 6 page 121.

Using circuit diagrams to make predictions

In this lesson you will use circuit diagrams to make predictions about whether the circuit will work.

Key words

buzzer
circuit diagram
symbol

Think back

Look at the circuit diagram.
How loud will the buzzers be?
How could you make them louder?

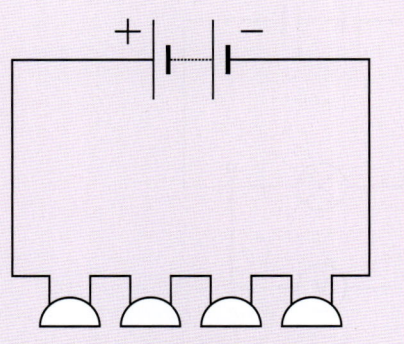

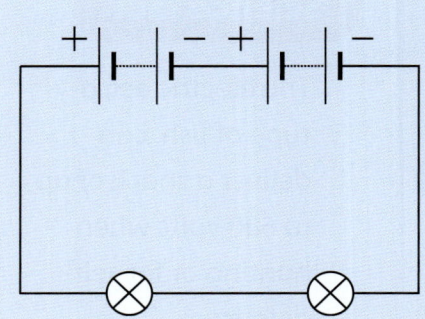

A student drew this diagram.

Will the circuit work? Can you explain why? Can you correct it?

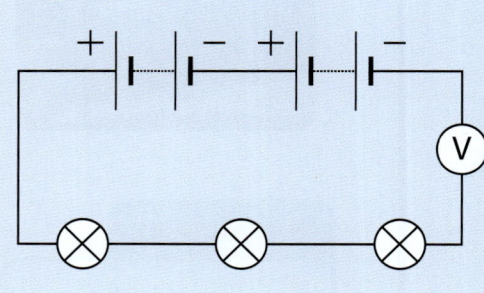

What is wrong with this student's diagram? Correct the diagram and explain why you need to change the circuit. Redraw it in your notebook.

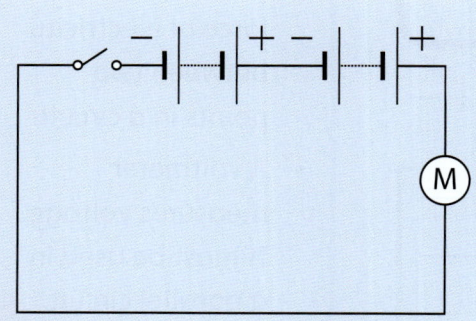

Another student drew this circuit. Will the motor work in this circuit?

What can you do to make it work?

■ For more activities, go to Workbook 6 page 122.

Revising circuit diagrams

1. In your notebook, draw an accurate circuit diagram with three batteries, wires, two bulbs and a closed switch. Colour in the bulb to show how bright you predict it will be.

2. In your notebook, draw another diagram with one battery, wires, two buzzers and an open switch. Will you hear the buzzer?

3. Your friend built this circuit but it does not work. How can you correct it?

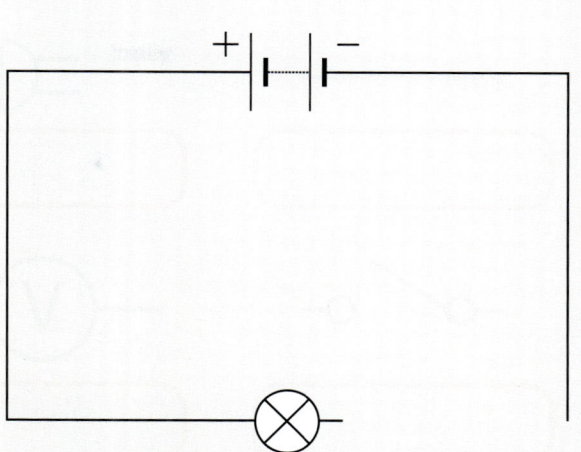

Working with circuit diagrams

1. Set up the circuit shown below.

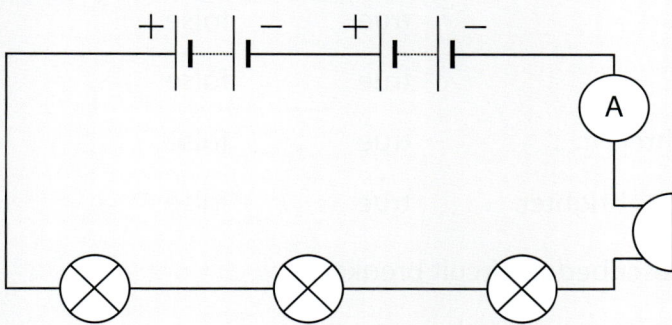

2. Predict what would happen if you removed each of the components in turn and re-connected the circuit so it is complete.

3. Carry out the investigation and record what happens.

Stretch zone

Design and make a set of traffic lights. Plan your circuit and then assemble the lights so they change when you operate a switch.

Key idea

Circuit diagrams can be used to predict if a circuit will work.

Check how much you know.
Try the questions on pages 124–125.

■ For more activities, go to Workbook 6 page 123.

1 Label the components shown below. Use the words in the word box.

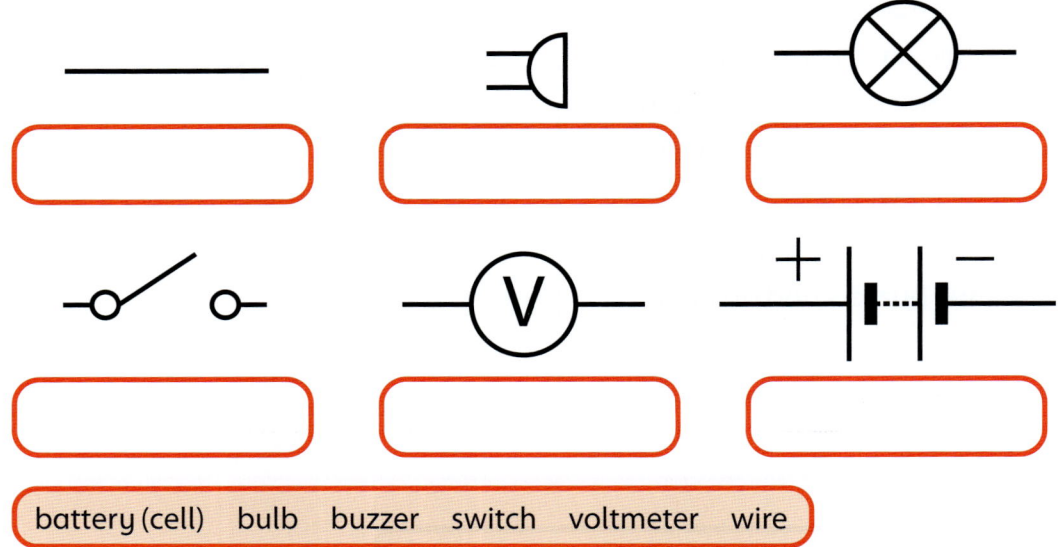

battery (cell)   bulb   buzzer   switch   voltmeter   wire

2 Circle whether the following statements are true or false.

Electricity travels through conductors.	true	false
Electricity travels through insulators.	true	false
Adding bulbs to a circuit makes them brighter.	true	false
Adding batteries to a circuit makes the bulbs brighter.	true	false

3 Circle any of the following that could be described as circuit breakers:

battery   buzzer   switch   wire

4 **a** Name three good conductors of electricity.

_____   _____   _____

b Explain why electrical wires are covered with plastic.

_____

5 Which one of the circuits below would allow the bulbs to light up? Tick the right box.

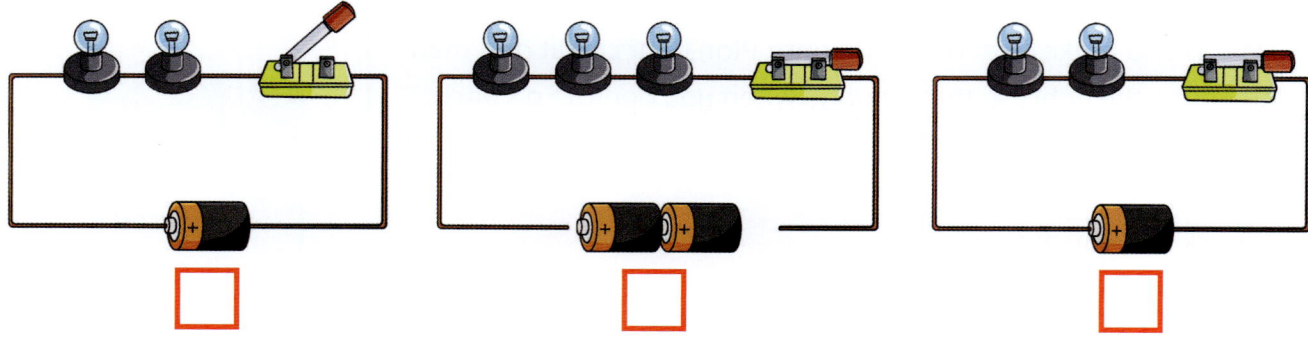

☐   ☐   ☐

■ For more activities, go to Workbook 6 page 124.

6 a Which of the following circuits would have the brightest bulbs? Give a reason.

A

B

I think circuit _____ will have the brightest bulbs.

This is because _____

_____

b How could you make the bulbs of circuit A brighter without altering the bulbs?

_____

7 Draw a circuit diagram for a circuit that contains a buzzer that can be turned on and off.

8 Explain why a voltmeter should be placed in a parallel circuit.

_____

9 A person investigates what happens to a circuit when the number of bulbs is changed. The person also changes the number of batteries.

a Is this a fair test?   yes   no

b Describe how you would carry out this test to make it a fair test.

_____

_____

■ For more activities, go to Workbook 6 page 125.

5 Adaptation and Inherited Characteristics

In this unit you will:

- recognise that living things have changed over time
- learn that fossils give us information about living things that lived on Earth millions of years ago
- review that living things produce offspring
- recognise that offspring vary and are not identical to the parents
- identify how animals and plants are adapted to suit their environment in different ways
- learn that adaptation may lead to new species of living things.

These animals eat leaves.

How are they adapted to their habitat?

Which ones will have problems when the lower leaves of the trees have all been eaten?

adaptation   ancestor
characteristic   environment
extinct   fossil   habitat
inherit   offspring
reproduction   species
variation

What is seen in the rock? Is it living, non-living or once living?

Discuss how the object got into the rock.

Can you see four moths?

They are the same species. How are they the same? How are they different?

Which moth would survive in a dark, polluted area?

Science fact

Scientists believe that 99% of all of the species that have ever lived on Earth have now died out. These are called extinct species. These include dinosaurs and the plant in the small photo above, which has not grown on Earth for over 200 million years.

■ For more activities, go to Workbook 6 pages 126–127.

The fossil record

In this lesson you will recognise that fossils provide information about living things that inhabited the Earth millions of years ago.

Key words

ancestor

cast

extinct

fossil

habitat

mould

Think back

What are fossils? Which type of rock are fossils found in? Why?

Fossils are evidence of living things preserved in rocks. The fossils can be parts of plants and animals turned to stone. These are called body fossils. They can also be old footprints. Sometimes fossils are made when an imprint of a living thing is filled up with material that hardens into rock. These are called trace fossils.

What information does this fossil give you about the type of animal and how it moved? Draw a picture to show what you think the living animal would have looked like.

Making a plaster cast fossil model

Choose an example of a once-living thing. This can be a seashell or parts of a plant.

1 Make a mould of your once-living thing by pressing it into modelling clay in a dish.

2 Pour some plaster of Paris into the dish and leave it to set for one hour.

3 Remove the plaster of Paris from the dish and leave it to dry overnight.

4 Paint and display your fossil in a class display. Move around the displays and try to identify the fossils.

Discuss how your model is the same as a fossil. How is it different?

Warning! Do not get plaster of Paris onto your skin or into your eyes. What could happen if you did?

■ For more activities, go to Workbook 6 page 128.

The plaster of Paris filled the mould and made a cast. Many fossils can be found as moulds or casts.

This dinosaur footprint was made 150 000 000 years ago

Is the footprint a mould or a cast?

Estimate the size of the footprint. How large do you think the animal would have been?

Fossils show that some types of living things have not changed in appearance for millions of years. Some examples are crocodiles and a plant called a horsetail.

Crocodile

Horsetail

Other living things have died out, so are not seen anymore. They are extinct.

Dinosaurs lived on the Earth between 65 and 245 000 000 years ago. Scientists think they became extinct after an asteroid hit the Earth and they could not adapt to the new conditions.

Stretch zone

Mary Anning was a famous fossil hunter. Write a short fact file about her. Include when and where she was born, how she became interested in fossils and details of her work.

Science fact

Scientists can find out a lot about ancient habitats by studying fossils. For example, finding fossil shells on Mount Everest proved that the mountain had been pushed up from under the sea.

Key idea

Fossils give us clues about the structure of living things, how they have changed, and what habitats were like in the past.

5 Adaptation and Inherited Characteristics

■ For more activities, go to Workbook 6 page 129.

Changes over time

In this lesson you will recognise that living things have changed over time.

Key words
ancestor
breed
characteristic
cross breeding
selective breeding

Study the fossil skeleton. Which modern animal does it look like? Has this type of animal changed much over the past 70 000 000 years?

Some types of animals and plants have not changed much over millions of years. Other living things have changed a lot.

These pictures show how scientists believe a horse's shape and leg structure has changed over millions of years. The evidence from this research is from fossil records and other detailed scientific dating methods, such as carbon dating.

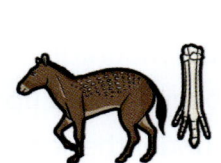

50 million years ago

35 million years ago

10 million years ago

5 million years ago

modern horse

Science fact

Some whales and dolphins have small bones in their fins. It is thought that these fins might have been back legs in their ancestors. Early whales and dolphins may have walked on land.

Discuss the ways in which modern horses are thought to be different from their ancestors. Why do you think these changes have happened?

■ For more activities, go to Workbook 6 page 130.

Handling data about change

1. Study the data below. It shows the average height of modern horses and their ancestors. It also shows when they lived.

Animal	Height in metres	Date the animals lived (millions of years ago)
A	0.4	45–55
B	0.6	30–40
C	1.0	15–25
D	1.25	5–15
E	1.6	present time–5

2. Present the data as a bar chart.

3. Identify any trends or patterns in the data.

4. Write a short report to explain how horses have changed over the past 55 000 000 years.

Why would you not use a line chart or a scatter chart for this data?

Plants have also changed over time. Some early grasses have changed to give modern cereals such as wheat.

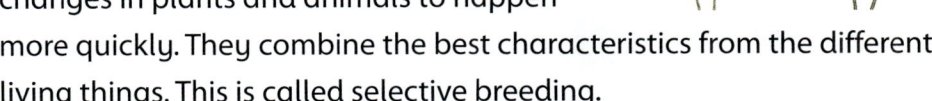

wild wheat   bread wheat

The mixing of characteristics can happen naturally, but farmers have been helping changes in plants and animals to happen more quickly. They combine the best characteristics from the different living things. This is called selective breeding.

Use your knowledge of plant life cycles to discuss why selective breeding is used on flowering plants but not non-flowering plants.

With plants, the pollen from one variety is added to the stigma of another so they cross-pollinate. This is called cross breeding or crossing.

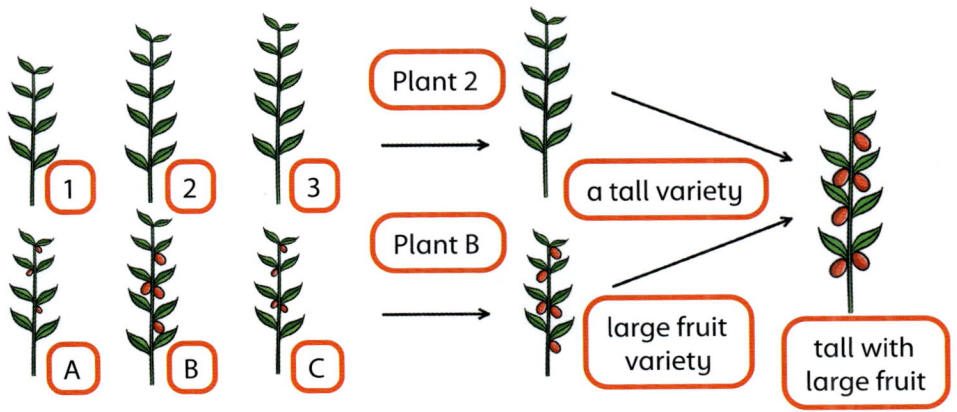

Plant 2 — a tall variety
Plant B — large fruit variety
tall with large fruit
1 2 3
A B C

Plants 2 and B have been crossed

Key ideas

- Many living things have changed over time and now do not look like their ancient ancestors.
- Farmers and scientists use selective breeding to produce more useful animal and plant varieties.

■ For more activities, go to Workbook 6 page 131.

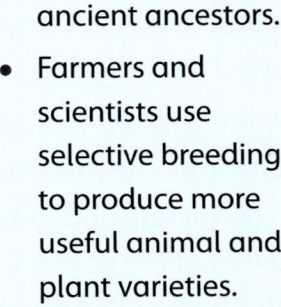

5 Adaptation and Inherited Characteristics

Offspring inherit characteristics

In this lesson you will recognise that living things produce offspring of the same kind.

Key words

characteristic
inherit
offspring
reproduce
species

Think back

Why do living things need to reproduce?

Tell your partner the name of an animal that has young that are not like the parents when born or hatched.

Individuals in each species cannot live forever. To make sure the species does not die out, living things must make new copies of themselves. These are called offspring. Many living things have offspring that look like smaller versions of the parents.

Study the animals in the photographs. Which of the young animals looks like their parent? What will happen to the other two animals to make them more like the parents?

Which animals will these offspring grow up to be? Do humans have offspring that look like their parents?

Plants can reproduce by making identical copies of themselves through making bulbs, budding or sending out runners. This is called asexual reproduction.

Flowering plants reproduce by a female ovule being fertilised by male pollen from the same or another plant. Seeds are produced and these can germinate and form new plants. The new plants are not identical to either of the parents, but have a mixture of characteristics from both. This is called sexual reproduction.

132

■ For more activities, go to Workbook 6 page 132.

The mixing of characteristics can lead to new varieties that are better adapted to a habitat or changing environment.

Seeds from a sunflower will grow and develop into a sunflower very like the parent plant.

Seed investigation

You will be given four different specimens of seeds. These will be labelled A, B, C and D.

1 Plan an investigation to find out which species of plant has produced each type of seed. Think carefully about the timings.

2 Research the types of seed to help you make a prediction. Then carry out your investigation.

3 Record your findings. Were your predictions correct?

4 Produce a short computer or poster presentation to describe your investigation and your findings.

Be a scientist

Scientists make sure they can follow every specimen throughout an investigation. They label everything very carefully.

▶ page 10

Did the plants produce offspring like the parent plant?

When a horse gives birth, the offspring will never be a camel or a giraffe. The parents pass on characteristics to the offspring that make them like the parents. We say that the offspring have inherited the characteristics from the parent.

These characteristics are mixed and so new variations can emerge, but each living thing only has offspring of the same type. Sunflower seeds will only give sunflower plants. Frog's eggs will only develop into tadpoles and then adult frogs. A butterfly egg will not hatch to give a moth or a dragonfly.

Science fact

Some animals, such as fish and insects, lay many eggs. This helps to make sure that at least some offspring will survive to become adults.

Butterfly eggs

Key ideas

- Living things need to reproduce to make offspring to prevent the species from becoming extinct.

- The offspring inherit characteristics from their parents.

Stretch zone

What might happen to a living organism if it inherits a characteristic that is not positive for its habitat?

■ For more activities, go to Workbook 6 page 133.

Variation in living things

In this lesson you will recognise that offspring vary and are not identical to their parents.

Key words
characteristic
class
offspring
reproduction
species
variation

Different classes of living things have very different characteristics from each other. It is these characteristics that help us to classify living things.

How are these animals different? How are they the same? Would you classify them all into the same close groupings?

Living things in the same species look very similar and are able to make offspring together. Individuals in a species do have small differences in characteristics. These are called variations. A person who keeps horses or goats can identify each individual animal from the group. No two individuals in a species are identical.

How might a person be able to tell different goats or camels apart from the group?

If you look around the room, you will see that there is variation in humans too. The chart below shows variation in height seen in adult humans. The shape of the chart is common when looking at variation.

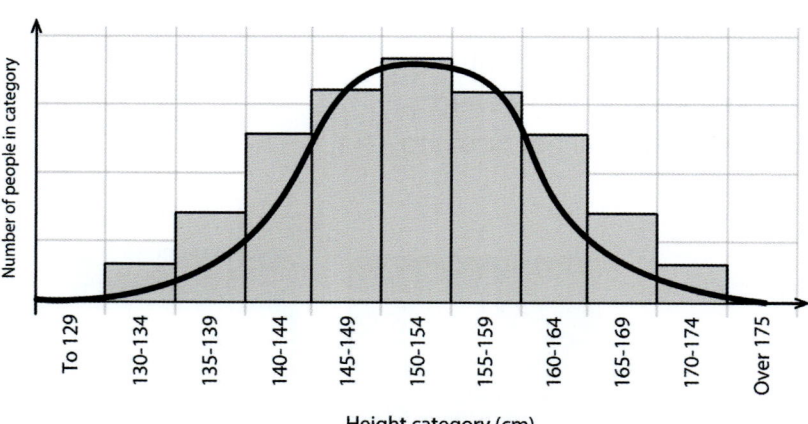

Height category (cm)

What does the chart tell you about the number of very tall or very short people? What is the average height for adult humans?

Be a scientist

The shape of charts can tell scientists a lot about data. This chart shows some values are at either end but most are gathered in the middle.

▶ page 11

134

Variation is natural and happens in most animal and plant species. Variation can be inherited. Inherited variation results from the mixing of information passed on from parents.

Variation can also be environmental. This is when changes happen because of how and where a living thing lives. For example, a lion that gets a lot of food will grow bigger than one that does not, and a plant living under a tree will grow taller to get more sunlight.

Investigating variation in shoe size

1 Plan a survey to find out if there is any variation in shoe size of people in your class.

2 Record the results in a table.

3 Produce a chart similar to the one opposite and identify any patterns in the results.

4 Produce a short information leaflet or blog to describe the variation in shoe size in your class.

What is the average shoe size for people in your class?

Is variation in shoe size inherited or environmental?

Environmental factors can affect plants a lot. For example, if one is kept out of sunlight or not given water and nutrients, it will look very different, as you can see in the photograph.

Investigating environmental variation in plants

1 Plan an investigation to explore variation in plants.

2 Collect seeds from a parent plant and observe and record how the seedlings develop.

3 Remember to plan a fair test and bear in mind environmental factors such as light and water.

4 Compare your plan with your group, and agree on the best way to carry out the investigation.

Stretch zone

Research how an environmental factor, such as shortage of food, can cause variation between the same species of animal. Present your ideas as a short talk.

Key ideas

- Individuals in a species are not identical. They show variation. This is natural.

- Variation can be inherited from parents and caused by the environment.

■ For more activities, go to Workbook 6 page 135.

Adapting to the environment

In this lesson you will identify how animals and plants are adapted to suit their environment in different ways.

Key words
adaptation
environment

Think back

How are polar bears adapted to live in cold areas?

How are cacti adapted to live in dry deserts?

Egret

Hummingbird

How are the egret and hummingbird adapted to their habitats?

Why could a hummingbird not wade to find and catch fish?

Animals and plants are adapted to live in their habitats. If they were not, they would not survive.

In animals, adaptations include body coverings, size, body shape and ways to catch food.

The Arctic fox eats rabbits. The rabbits have white fur in the winter and brown fur in the summer.

The fur on the Arctic fox keeps it warm in a cold habitat

How else does the fur help the Arctic fox fit into its habitat?

Why does the rabbit fur colour change?

Designer animal

You are going to design a model of an animal that is adapted to the habitat described opposite. Like a good scientist, you will need to use your imagination and creativity.

1 List the useful characteristics your animal will need to live in its habitat.

2 Consider how it will move, how it will catch food, and how it will stop from overheating.

3 Draw your animal and label the characteristics.

4 Present your design in a class display.

The habitat is very hot and dry, but floods regularly. The land has a lot of soft sand. The animal feeds on plants but can also eat insects and small birds.

136

■ For more activities, go to Workbook 6 page 136.

Plant adaptations include leaf shape and size, flower designs, ways to prevent drying out and ways to stop being eaten.

Why do you think conifers are shaped in this way?

Conifers

Conifers are adapted to live in cold areas. They have a thick bark and grow close together to protect against the cold. They have thick leaves or spines covered in wax to prevent loss of water. The trees become narrower at the top and have branches pointing downwards.

Investigating adaptations of local plants

1 In a group, plan a survey of your local area to study plant adaptations.

2 Look for any ways that the plants have shapes to help them to survive.

3 Study the shape, height, leaves, flowers and roots, if possible.

4 Produce a presentation of your findings to share with the class.

Be a scientist

Scientists research their topic using books and the internet before they start a survey or investigation.

▶ page 8

Being well adapted to a habitat can have disadvantages. A living thing will not be adapted to other habitats. A camel could not live on a coral reef and a conifer could not live in the desert. Also, if a habitat changes then the living thing might not be able to survive.

Key idea

Living things are successful if they are adapted to live in their habitat. However, they may not survive if their habitat changes.

Stretch zone

Research how shellfish such as limpets from the same species are adapted differently to live in sheltered coasts or coasts that are hit by waves. Produce a poster to display your research.

■ For more activities, go to Workbook 6 page 137.

Survival and change

In this lesson you will identify how adaptations can help living things to survive and change.

Key words
adaptation
survival
variation

Think back

How are conifers and Arctic foxes adapted to live in their habitats?

How are the moths adapted to their habitat? Which moth is the most likely to be found and eaten?

Do you remember these moths?

If every individual in a species is identical, they will all be easily found by predators or they will all catch a disease. This could make the whole species extinct. Because individuals vary, it is more likely that some will survive.

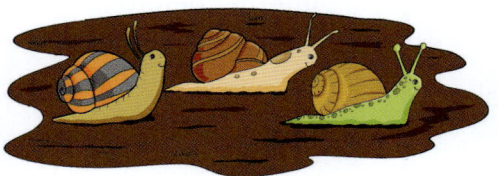

Which coloured variety of snail is most likely to survive to have offspring in a dark area? Why?

Plants show variation as well as animals. If a tree in a rainforest has a smoother bark than other trees of the same type it might survive better. This is because climbing plants that grow and smother trees will not be able to grip onto the trunk.

Change is common in most habitats. There are daily and seasonal changes. Unexpected changes such as drought, flooding and new predators arriving also happen.

Having a characteristic that helps an individual survive when others die out is called 'survival of the fittest'. Only the living things that survive go on to have offspring, which have the useful characteristic.

138

■ For more activities, go to Workbook 6 page 138.

Adaptation trail

You are going to set out an adaptation trail.

1. Take some pieces of coloured card and cut out 15 identical insect shapes. Five should be blue, five should be green and five should be red. This is your population.

2. Place your insect shapes on plants around the school.

3. Ask someone to look for your insects for five minutes and 'catch' any they see.

4. Collect any insects they have not found.

5. Make five more of any insects that were not found. What does your population look like now?

6. Make a poster to share your ideas about adaptation and how populations can change.

How does the change in your population explain 'survival of the fittest'?

Living things that are well adapted to their habitat survive and pass the adaptations on from generation to generation. Over time, this process can lead to big changes in the living things. For example, it could lead to longer legs in wading birds or plants that flower earlier than other plants.

There can be so much change that new species develop. Remember the finches from page 19? A few finches were blown from the mainland to some of the Galapagos islands. Over time they adapted to be able to eat different foods. Now they have changed into different species.

Using secondary sources of information

1. Research why the dodo became extinct.

2. Produce a short presentation to share with the class.

 Stretch zone

Research a plant that became extinct because it was not able to adapt to its changing surroundings. Explain your findings to a partner.

Check how much you know.
Try the questions on pages 140–141.

Key ideas

- Variation makes it more likely for individuals to survive if conditions in a habitat change.
- Variation can eventually result in a new species.

■ For more activities, go to Workbook 6 page 139.

What have I learned about adaptation and inherited characteristics?

1

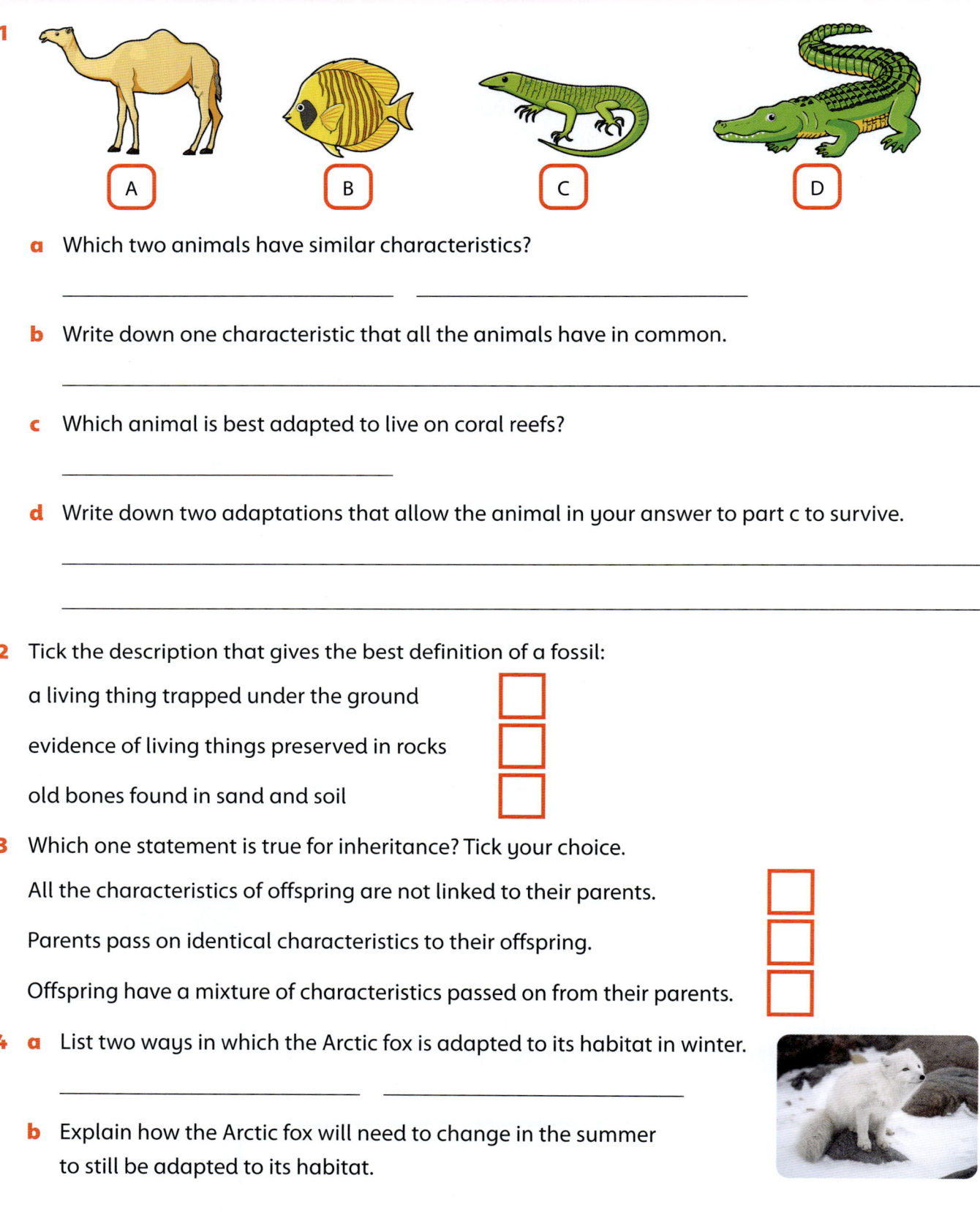

A   B   C   D

a Which two animals have similar characteristics?

_____   _____

b Write down one characteristic that all the animals have in common.

_____

c Which animal is best adapted to live on coral reefs?

_____

d Write down two adaptations that allow the animal in your answer to part c to survive.

_____

_____

2 Tick the description that gives the best definition of a fossil:

a living thing trapped under the ground ☐

evidence of living things preserved in rocks ☐

old bones found in sand and soil ☐

3 Which one statement is true for inheritance? Tick your choice.

All the characteristics of offspring are not linked to their parents. ☐

Parents pass on identical characteristics to their offspring. ☐

Offspring have a mixture of characteristics passed on from their parents. ☐

4 a List two ways in which the Arctic fox is adapted to its habitat in winter.

_____   _____

b Explain how the Arctic fox will need to change in the summer to still be adapted to its habitat.

_____

_____

■ For more activities, go to Workbook 6 page 140.

5

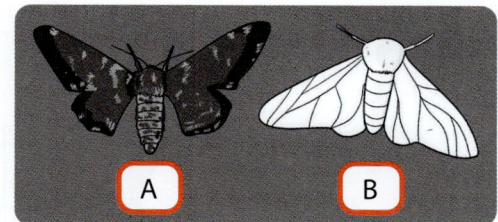

Which of the moths above is most likely to survive and have offspring? Why?

_____

_____

6 Look at the pictures and study the table.

| 50 million years ago | 35 million years ago | 10 million years ago | 5 million years ago | modern horse |

The data shows the heights of the different ancestors of a modern horse.

Ancestor	Date the animal lived (millions of years ago)	Height in metres
A	45–55	0.4
B	30–40	0.6
C	15–25	1.0
D	5–15	1.25
E	present–5	1.6

a What does the data tell you about the heights of horse ancestors over time?

_____

b Calculate how much the height of horses and their ancestors has changed over 55 million years.

_____

_____

c Write down **one** advantage to horses of this change in height.

_____

d Give an example of a modern animal that has not changed over millions of years.

_____

■ For more activities, go to Workbook 6 page 141.

Glossary

adaptation

ammeter

ancestor

battery

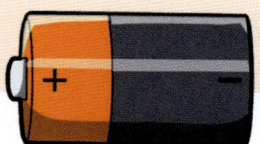

beam

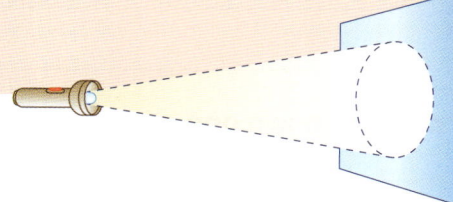

bulb

buzzer

characteristic

circuit diagram

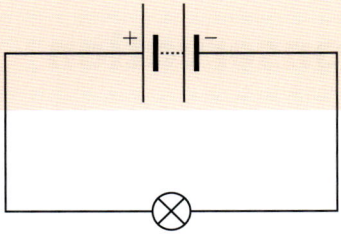

circulatory system

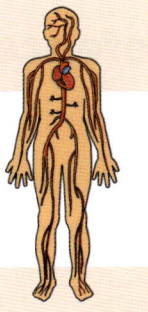

classification

component

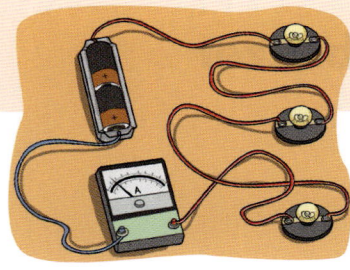

conductor

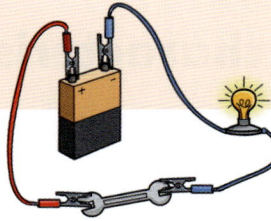

conservation

defence mechanism

deforestation

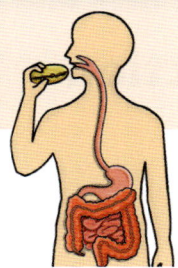

digestive system

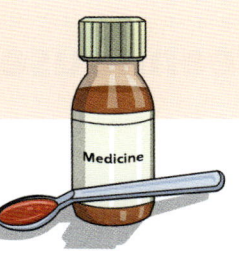

drug

environment

extinct

fossil

function

greenhouse effect

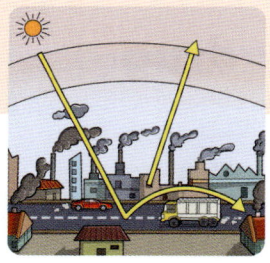

habitat

infectious disease

inherit

insulator

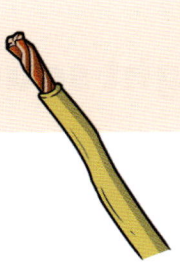

key

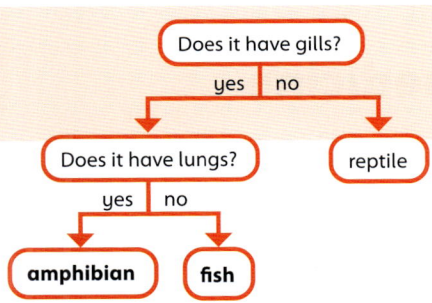

kingdom

lifestyle

light intensity

light source

medicine

microorganism

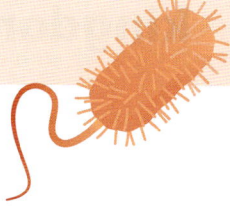

mirror

nervous system

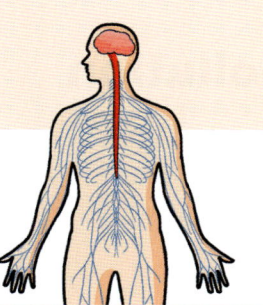

offspring

opaque

organ

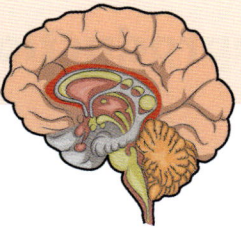

parallel circuit

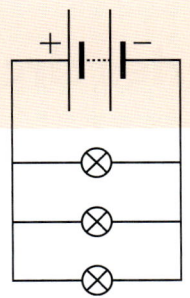

pollution

ray

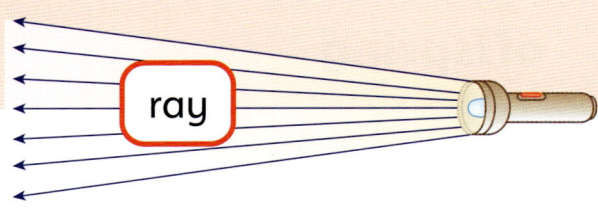

reflect

reproduction

series circuit

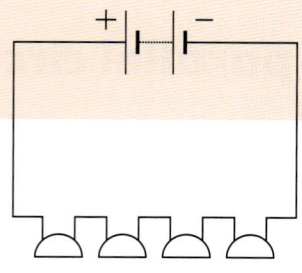

shadow

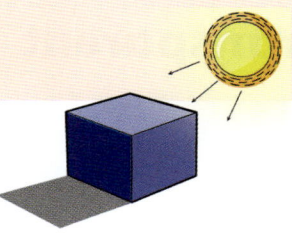

silhouette

species

switch

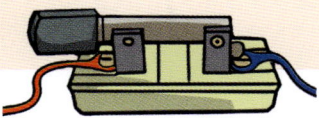

translucent

transparent

urinary system

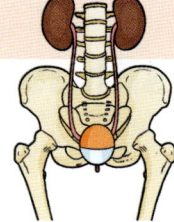

vaccine

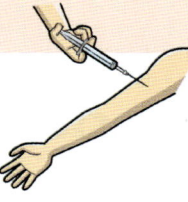

variation

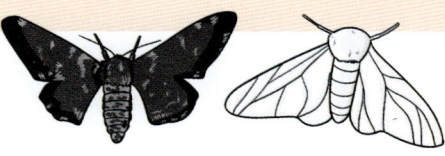

voltage

voltmeter